CISM®
Certified Information Security Manager®
An ISACA® Certification

CERTIFIED INFORMATION SECURITY MANAGER®

CISM® Review Questions, Answers & Explanations Manual 2012

ISACA®
Trust in, and value from, information systems

ISACA®

With 95,000 constituents in 160 countries, ISACA (*www.isaca.org*) is a leading global provider of knowledge, certifications, community, advocacy and education on information systems (IS) assurance and security, enterprise governance and management of IT, and IT-related risk and compliance. Founded in 1969, the nonprofit, independent ISACA hosts international conferences, publishes the *ISACA® Journal*, and develops international IS auditing and control standards, which help its constituents ensure trust in, and value from, information systems. It also advances and attests IT skills and knowledge through the globally respected Certified Information Systems Auditor® (CISA®), Certified Information Security Manager® (CISM®), Certified in the Governance of Enterprise IT® (CGEIT®) and Certified in Risk and Information Systems Control™ (CRISC™) designations. ISACA continually updates COBIT®, which helps IT professionals and enterprise leaders fulfill their IT governance and management responsibilities, particularly in the areas of assurance, security, risk and control, and deliver value to the business.

Disclaimer

ISACA has designed and created *CISM® Questions, Answers & Explanations Manual 2012* (the "Work") primarily as an educational resource to assist individuals preparing to take the CISM certification exam. It was produced independently from the CISM exam and the CISM Certification Committee, which has had no responsibility for its content. Copies of past exams are not released to the public and were not made available to ISACA for preparation of this publication. ISACA makes no representations or warranties whatsoever with regard to these or other ISACA publications assuring candidates' passage of the CISM exam.

Reservation of Rights

ISACA

3701 Algonquin Road, Suite 1010
Rolling Meadows, Illinois 60008 USA
Phone: +1.847.253.1545
Fax: +1.847.253.1443
E-mail: *info@isaca.org*
Web site: *www.isaca.org*

ISBN 978-1-60420-216-8
CISM® Review Questions, Answers & Explanations Manual 2012
Printed in the United States of America

PREFACE

ISACA is pleased to offer this *CISM® Review Questions, Answers & Explanations Manual 2012*. The purpose of this manual is to provide the CISM candidate with sample questions and testing topics to help prepare and study for the CISM exam.

The material in this manual consists of 700 multiple-choice study questions, answers and explanations, which are organized according to the newly revised (for 2012) CISM job practice domains. The questions in this manual appeared in the *CISM® Review Questions, Answers & Explanations Manual 2011* and in the *CISM® Review Questions, Answers & Explanations Manual 2011 Supplement* and have been reorganized to reflect the job practice that is effective in 2012. These questions, answers and explanations are intended to introduce the CISM candidate to the types of questions that appear on the CISM exam. They are not actual questions from the exam. Questions are sorted by CISM job practice domains and a sample exam of 200 questions is also provided. Sample questions contained in this manual are provided to assist the CISM candidate in understanding the material in the *CISM® Review Manual 2012* and to depict the type of question format typically found on the CISM exam.

A job practice study is conducted at least every five years to ensure that the CISM certification is current and relevant. Further details regarding the new job practice can be found in the section on page v titled New—CISM Job Practice.

ISACA wishes you success with the CISM exam. Your commitment to pursuing the leading certification for information security managers is exemplary, and we welcome your comments and suggestions on the use and coverage of this manual. After taking the exam, please take a moment to complete the online questionnaire (*www.isaca.org/studyaidsevaluation*). Your observations will be invaluable as new questions, answers and explanations are prepared.

ACKNOWLEDGMENTS

The *CISM® Review Questions, Answers & Explanations Manual 2012* is the result of the collective efforts of many volunteers over the past several years. ISACA members from throughout the global IS audit and control profession participated, generously offering their talents and expertise. This international team exhibited a spirit and selflessness that has become the hallmark of contributors to this valuable manual. Their participation and insight are truly appreciated.

NEW—CISM JOB PRACTICE

BEGINNING IN 2012, THE CISM EXAM WILL TEST THE NEW CISM JOB PRACTICE.

An international job practice analysis is conducted at least every five years or sooner to maintain the validity of the CISM certification program. A new job practice forms the basis of the CISM exam beginning in June 2012.

The primary focus of the job practice is on the current tasks performed and the knowledge used by CISMs. By gathering evidence of the current work practice of CISMs, ISACA is able to ensure that the CISM program continues to meet the high standards for the certification of professionals throughout the world.

The findings of the CISM job practice analysis are carefully considered and directly influence the development of new test specifications to ensure that the CISM exam reflects the most current best practices.

The new 2012 job practice reflects the areas of study to be tested and is compared below to the previous job practice.

Previous CISM Job Practice	New 2012 CISM Job Practice
Domain 1: Information Security Governance (23%) Domain 2: Information Risk Management (22%) Domain 3: Information Security Program Development (17%) Domain 4: Information Security Program Management (24%) Domain 5: Incident Management and Response (14%)	Domain 1: Information Security Governance (24%) Domain 2: Information Risk Management and Compliance (33%) Domain 3: Information Security Program Development and Management (25%) Domain 4: Information Security Incident Management (18%)

TABLE OF CONTENTS

INTRODUCTION

OVERVIEW

This manual consists of 700 multiple-choice questions, answers and explanations (numbered S1-1, S1-2, etc.). These questions are selected and provided in two formats.

Questions Sorted by Domain

Questions, answers and explanations are provided (sorted) by the CISM job practice domains. This allows the CISM candidate to refer to specific questions to evaluate comprehension of the topics covered within each domain. These questions are representative of CISM questions, although they are not actual exam items. They are provided to assist the CISM candidate in understanding the material in the *CISM® Review Manual 2012* and to depict the type of question format typically found on the CISM exam. The numbers of questions, answers and explanations provided in the four domain chapters in this publication provide the CISM candidate with a maximum number of study questions.

Sample Exam

A random sample exam of 200 of the questions is also provided in this manual. **This exam is organized according to the domain percentages specified in the CISM job practice and used on the CISM exam:**

Information Security Governance.. 24 percent
Information Risk Management and Compliance.. 33 percent
Information Security Program Development and Management........................... 25 percent
Information Security Incident Management... 18 percent

Candidates are urged to use this sample exam and the answer sheets provided to simulate an actual exam. Many candidates use this exam as a pretest to determine strengths or weaknesses, or as a final exam. Sample exam answer sheets have been provided for both uses. In addition, a sample exam answer/reference key is included. These sample exam questions have been cross-referenced to the questions, answers and explanations by domain, so it is convenient to refer to the explanations of the correct answers. This publication is ideal to use in conjunction with the *CISM® Review Manual 2012*, and with the *CISM® Review Questions, Answers & Explanations Manual 2012 Supplement*.

It should be noted that the *CISM® Review Questions, Answers & Explanations Manual 2012* has been developed to assist the CISM candidate in studying and preparing for the CISM exam. As you use this publication to prepare for the exam, please note that it covers a broad spectrum of information security management issues. Do not assume that reading and working the questions in this manual will fully prepare you for the exam. Since exam questions often relate to practical experience, the CISM candidate is cautioned to refer to his/her own experience and to other publications referred to in the *CISM® Review Manual 2012*. These additional references are an excellent source of further detailed information and clarification. It is recommended that candidates evaluate the job practice domains in which he/she feels weak or requires a further understanding, and study accordingly.

Also, please note that this publication has been written using standard American English.

TYPES OF QUESTIONS ON THE CISM EXAM

CISM exam questions are developed with the intent of measuring and testing practical knowledge, and the application of information security managerial principles and standards. As previously mentioned, all questions are presented in a multiple choice format and are designed for one best answer.

The candidate is cautioned to read each question carefully. Many times a CISM exam question will require the candidate to choose the appropriate answer that is **MOST** likely or **BEST**, or, the candidate may be asked to choose a practice or procedure that would be performed **FIRST** related to the other answers. In every case, the candidate is required to read the question carefully, eliminate known wrong answers and then make the best choice possible. Knowing that these types of questions are asked and how to study to answer them will go a long way toward answering them correctly.

Each CISM question has a stem (question) and four options (answer choices). The candidate is asked to choose the correct or best answer from the options. The stem may be in the form of a question or incomplete statement. In some instances, a scenario or description also may be included. These questions normally include a description of a situation and require the candidate to answer two or more questions based on the information provided.

Another condition the candidate should consider when preparing for the exam is to recognize that information security is a global profession, and individual perceptions and experiences may not reflect the more global position or circumstance. Since the exam and CISM manuals are written for the international information security community, the candidate will be required to be somewhat flexible when reading a condition that may be contrary to the candidate's experience. It should be noted that CISM exam questions are written by experienced information security managers from around the world. Each question on the exam is reviewed by ISACA's CISM Test Enhancement Subcommittee and CISM Certification Committee, which consist of international members. This geographic representation ensures that all exam questions are understood equally in every country and language.

> **Note:** ISACA review manuals are living documents. As technology advances, ISACA manuals will be updated to reflect such advances. Further updates to this document before the date of the exam may be viewed at *www.isaca.org/studyaidupdates*.

Any suggestions to enhance the materials covered herein, or reference materials, should be submitted online at *www.isaca.org/studyaidsevaluation*.

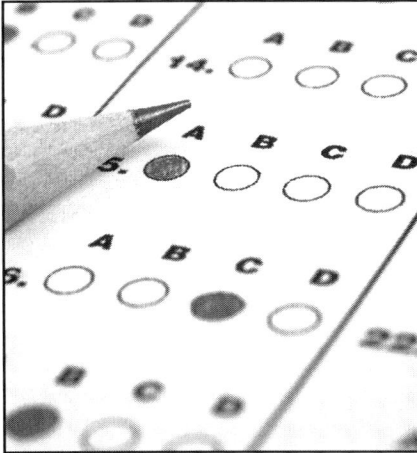

PRETEST

If you wish to take a pretest to determine strengths and weaknesses, the Sample Exam begins on page 195 and the pretest answer sheet begins on page 227. You can score your pretest with the Sample Exam Answer and Reference Key on page 225.

Page intentionally left blank

QUESTIONS, ANSWERS AND EXPLANATIONS BY DOMAIN

DOMAIN 1—INFORMATION SECURITY GOVERNANCE (24%)

S1-1   Which of the following should be the **FIRST** step in developing an information security plan?

A. Perform a technical vulnerabilities assessment
B. Analyze the current business strategy
C. Perform a business impact analysis
D. Assess the current levels of security awareness

B   Prior to assessing technical vulnerabilities or levels of security awareness, an information security manager needs to gain an understanding of the current business strategy and direction. A business impact analysis should be performed prior to developing a business continuity plan, but this would not be an appropriate first step in developing an information security strategy because it focuses on availability.

S1-2   Senior management commitment and support for information security can **BEST** be obtained through presentations that:

A. use illustrative examples of successful attacks.
B. explain the technical risks to the organization.
C. evaluate the organization against best security practices.
D. tie security risks to key business objectives.

D   Senior management seeks to understand the business justification for investing in security. This can best be accomplished by tying security to key business objectives. Senior management will not be as interested in technical risks or examples of successful attacks if they are not tied to the impact on business environment and objectives. Industry best practices are important to senior management but, again, senior management will give them the right level of importance when they are presented in terms of key business objectives.

S1-3   The **MOST** appropriate role for senior management in supporting information security is the:

A. evaluation of vendors offering security products.
B. assessment of risks to the organization.
C. approval of policy statements and funding.
D. monitoring adherence to regulatory requirements.

C   Since the members of senior management are ultimately responsible for information security, they are the ultimate decision makers in terms of governance and direction. They are responsible for approval of major policy statements and requests to fund the information security practice. Evaluation of vendors, assessment of risks and monitoring compliance with regulatory requirements are day-to-day responsibilities of the information security manager; in some organizations, business management is involved in these other activities, though their primary role is direction and governance.

S1-4   Which of the following would **BEST** ensure the success of information security governance within an organization?

        A. Steering committees approve security projects
        B. Security policy training provided to all managers
        C. Security training available to all employees on the intranet
        D. Steering committees enforce compliance with laws and regulations

A   The existence of a steering committee that approves all security projects would be an indication of the existence of a good governance program. Compliance with laws and regulations is part of the responsibility of the steering committee but it is not a full answer. Awareness training is important at all levels in any medium, and also an indicator of good governance. However, it must be guided and approved as a security project by the steering committee.

S1-5   Information security governance is **PRIMARILY** driven by:

        A. technology constraints.
        B. regulatory requirements.
        C. litigation potential.
        D. business strategy.

D   Governance is directly tied to the strategy and direction of the business. Technology constraints, regulatory requirements and litigation potential are all important factors, but they are necessarily in line with the business strategy.

S1-6   Which of the following represents the **MAJOR** focus of privacy regulations?

        A. Unrestricted data mining
        B. Identity theft
        C. Human rights protection
        D. Identifiable personal data

D   Protection of identifiable personal data is the major focus of privacy regulations such as the Health Insurance Portability and Accountability Act (HIPAA). Data mining is an accepted tool for ad hoc reporting; it could pose a threat to privacy only if it violates regulatory provisions. Identity theft is a potential consequence of privacy violations but not the main focus of many regulations. Human rights addresses privacy issues but is not the main focus of regulations.

S1-7   Investments in information security technologies should be based on:

        A. vulnerability assessments.
        B. value analysis.
        C. business climate.
        D. audit recommendations.

B   Investments in security technologies should be based on a value analysis and a sound business case. Demonstrated value takes precedence over the current business climate because it is ever changing. Basing decisions on audit recommendations would be reactive in nature and might not address the key business needs comprehensively. Vulnerability assessments are useful, but they do not determine whether the cost is justified.

S1-8   Retention of business records should **PRIMARILY** be based on:

      A. business strategy and direction.
      B. regulatory and legal requirements.
      C. storage capacity and longevity.
      D. business case and value analysis.

B   Retention of business records is generally driven by legal and regulatory requirements. Business strategy and direction would not normally apply nor would they override legal and regulatory requirements. Storage capacity and longevity are important but secondary issues. Business case and value analysis would be secondary to complying with legal and regulatory requirements.

S1-9   Which of the following is characteristic of centralized information security management?

      A. More expensive to administer
      B. Better adherence to policies
      C. More aligned with business unit needs
      D. Faster turnaround of requests

B   Centralization of information security management results in greater uniformity and better adherence to security policies. It is generally less expensive to administer due to the economies of scale. However, turnaround can be slower due to the lack of alignment with business units.

S1-10   Successful implementation of information security governance will **FIRST** require:

      A. security awareness training.
      B. updated security policies.
      C. a computer incident management team.
      D. a security architecture.

B   Updated security policies are required to align management objectives with security procedures; management objectives translate into policy, policy translates into procedures. Security procedures will necessitate specialized teams such as the computer incident response and management group as well as specialized tools such as the security mechanisms that comprise the security architecture. Security awareness will promote the policies, procedures and appropriate use of the security mechanisms.

S1-11   Which of the following individuals would be in the **BEST** position to sponsor the creation of an information security steering group?

      A. Information security manager
      B. Chief operating officer (COO)
      C. Internal auditor
      D. Legal counsel

B   The chief operating officer (COO) is highly-placed within an organization and has the most knowledge of business operations and objectives. The chief internal auditor and chief legal counsel are appropriate members of such a steering group. However, sponsoring the creation of the steering committee should be initiated by someone versed in the strategy and direction of the business. Since a security manager is looking to this group for direction, they are not in the best position to oversee formation of this group.

S1-12   The **MOST** important component of a privacy policy is:

        A. notifications.
        B. warranties.
        C. liabilities.
        D. geographic coverage.

A   Privacy policies must contain notifications and opt-out provisions; they are a high-level management statement of direction. They do not necessarily address warranties, liabilities or geographic coverage, which are more specific.

S1-13   The cost of implementing a security control should not exceed the:

        A. annualized loss expectancy.
        B. cost of an incident.
        C. asset value.
        D. implementation opportunity costs.

C   The cost of implementing security controls should not exceed the worth of the asset. Annualized loss expectancy represents the losses that are expected to happen during a single calendar year. A security mechanism may cost more than this amount (or the cost of a single incident) and still be considered cost effective. Opportunity costs relate to revenue lost by forgoing the acquisition of an item or the making of a business decision.

S1-14   Which of the following is **MOST** appropriate for inclusion in an information security strategy?

        A. Business controls designated as key controls
        B. Security processes, methods, tools and techniques
        C. Firewall rule sets, network defaults and intrusion detection system (IDS) settings
        D. Budget estimates to acquire specific security tools

B   A set of security objectives, processes, methods, tools and techniques together constitute a security strategy. Although IT and business governance are intertwined, business controls may not be included in a security strategy. Budgets will generally not be included in an information security strategy. Additionally, until information security strategy is formulated and implemented, specific tools will not be identified and specific cost estimates will not be available. Firewall rule sets, network defaults and intrusion detection system (IDS) settings are technical details subject to periodic change, and are not appropriate content for a strategy document.

S1-15   Senior management commitment and support for information security will **BEST** be attained by an information security manager by emphasizing:

        A. organizational risk.
        B. organizationwide metrics.
        C. security needs.
        D. the responsibilities of organizational units.

A   Information security exists to help the organization meet its objectives. The information security manager should identify information security needs based on organizational needs. Organizational or business risk should always take precedence. Involving each organizational unit in information security and establishing metrics to measure success will be viewed favorably by senior management after the overall organizational risk is identified.

S1-16   Which of the following roles would represent a conflict of interest for an information security manager?

     A. Evaluation of third parties requesting connectivity
     B. Assessment of the adequacy of disaster recovery plans
     C. Final approval of information security policies
     D. Monitoring adherence to physical security controls

C   Since management is ultimately responsible for information security, it should approve information security policy statements; the information security manager should not have final approval. Evaluation of third parties requesting access, assessment of disaster recovery plans and monitoring of compliance with physical security controls are acceptable practices and do not present any conflicts of interest.

S1-17   Which of the following situations must be corrected **FIRST** to ensure successful information security governance within an organization?

     A. The information security department has difficulty filling vacancies.
     B. The chief information officer (CIO) approves security policy changes.
     C. The information security oversight committee only meets quarterly.
     D. The data center manager has final signoff on all security projects.

D   A steering committee should be in place to approve all security projects. The fact that the data center manager has final signoff for all security projects indicates that a steering committee is not being used and that information security is relegated to a subordinate place in the organization. This would indicate a failure of information security governance. It is not inappropriate for an oversight or steering committee to meet quarterly. Similarly, it may be desirable to have the chief information officer (CIO) approve the security policy due to the size of the organization and frequency of updates. Difficulty in filling vacancies is not uncommon due to the shortage of good, qualified information security professionals.

S1-18   Which of the following requirements would have the lowest level of priority in information security?

     A. Technical
     B. Regulatory
     C. Privacy
     D. Business

A   Information security priorities may, at times, override technical specifications, which then must be rewritten to conform to minimum security standards. Regulatory and privacy requirements are government-mandated and, therefore, not subject to override. The needs of the business should always take precedence in deciding information security priorities.

S1-19   When an organization hires a new information security manager, which of the following goals should this individual pursue **FIRST**?

     A. Develop a security architecture
     B. Establish a steering committee
     C. Assemble an experienced staff
     D. Benchmark peer organizations

B   New information security managers should seek to build rapport and establish lines of communication with senior management to enlist their support. This is best accomplished by forming an information security steering committee. Benchmarking peer organizations is beneficial to better understand industry best practices, but it is secondary to obtaining senior management support. Similarly, developing a security architecture and assembling an experienced staff are objectives that can be obtained later.

S1-20   Which of the following is **MOST** likely to be discretionary?

    A. Policies
    B. Procedures
    C. Guidelines
    D. Standards

C   Policies define security goals and expectations for an organization. These are defined in more specific terms within standards and procedures. Standards establish what is to be done while procedures describe how it is to be done. Guidelines provide recommendations that business management must consider in developing practices within their areas of control; as such, they are discretionary.

S1-21   Security technologies should be selected **PRIMARILY** on the basis of their:

    A. ability to mitigate business risks.
    B. evaluations in trade publications.
    C. use of new and emerging technologies.
    D. benefits in comparison to their costs.

A   The most fundamental evaluation criterion for the appropriate selection of any security technology is its ability to reduce or eliminate business risks. Investments in security technologies should be based on their overall value in relation to their cost; the value can be demonstrated in terms of risk mitigation. This should take precedence over whether they use new or exotic technologies or how they are evaluated in trade publications.

S1-22   Which of the following are seldom changed in response to technological changes?

    A. Standards
    B. Procedures
    C. Policies
    D. Guidelines

C   Policies are high-level statements of objectives. Because of their high-level nature and statement of broad operating principles, they are less subject to periodic change. Security standards and procedures as well as guidelines must be revised and updated based on the impact of technology changes.

S1-23   The **MOST** important factor in planning for the long-term retention of electronically stored business records is to take into account potential changes in:

    A. storage capacity and shelf life.
    B. regulatory and legal requirements.
    C. business strategy and direction.
    D. application systems and media.

D   Long-term retention of business records may be severely impacted by changes in application systems and media. For example, data stored in nonstandard formats that can only be read and interpreted by previously decommissioned applications may be difficult, if not impossible, to recover. Business strategy and direction do not generally apply, nor do legal and regulatory requirements. Storage capacity and shelf life are important but secondary issues.

S1-24   Which of the following is characteristic of decentralized information security management across a
         geographically dispersed organization?

         A. More uniformity in quality of service
         B. Better adherence to policies
         C. Better alignment to business unit needs
         D. More savings in total operating costs

C   Decentralization of information security management generally results in better alignment to business unit
         needs. It is generally more expensive to administer due to the lack of economies of scale. Uniformity in
         quality of service tends to vary from unit to unit.

S1-25   Which of the following is the **MOST** appropriate position to sponsor the design and implementation of a
         new security infrastructure in a large global enterprise?

         A. Chief security officer (CSO)
         B. Chief operating officer (COO)
         C. Chief privacy officer (CPO)
         D. Chief legal counsel (CLC)

B   The chief operating officer (COO) is most knowledgeable of business operations and objectives. The
         chief privacy officer (CPO) and the chief legal counsel (CLC) may not have the knowledge of the
         day-to-day business operations to ensure proper guidance, although they have the same influence within
         the organization as the COO. Although the chief security officer (CSO) is knowledgeable of what is
         needed, the sponsor for this task should be someone with far-reaching influence across the organization.

S1-26   Which of the following would be the **MOST** important goal of an information security governance program?

         A. Review of internal control mechanisms
         B. Effective involvement in business decision making
         C. Total elimination of risk factors
         D. Ensuring trust in data

D   The development of trust in the integrity of information among stakeholders should be the primary goal
         of information security governance. Review of internal control mechanisms relates more to auditing,
         while the total elimination of risk factors is not practical or possible. Proactive involvement in business
         decision making implies that security needs dictate business needs when, in fact, just the opposite is true.
         Involvement in decision making is important only to ensure business data integrity so that data can
         be trusted.

S1-27   Acceptable levels of information security risk should be determined by:

         A. legal counsel.
         B. security management.
         C. external auditors.
         D. the steering committee.

D   Senior management, represented in the steering committee, has ultimate responsibility for determining
         what levels of risk the organization is willing to assume. Legal counsel, the external auditors and security
         management are not in a position to make such a decision.

S1-28   The **PRIMARY** goal in developing an information security strategy is to:

A.   establish security metrics and performance monitoring.
B.   educate business process owners regarding their duties.
C.   ensure that legal and regulatory requirements are met.
D.   support the business objectives of the organization.

D   The business objectives of the organization supersede all other factors. Establishing metrics and measuring performance, meeting legal and regulatory requirements, and educating business process owners are all subordinate to this overall goal.

S1-29   Senior management commitment and support for information security can **BEST** be enhanced through:

A.   a formal security policy sponsored by the chief executive officer (CEO).
B.   regular security awareness training for employees.
C.   periodic review of alignment with business management goals.
D.   senior management signoff on the information security strategy.

C   Ensuring that security activities continue to be aligned and support business goals is critical to obtaining their support. Although having the chief executive officer (CEO) signoff on the security policy and senior management signoff on the security strategy makes for good visibility and demonstrates good tone at the top, it is a one-time discrete event that may be quickly forgotten by senior management. Security awareness training for employees will not have as much effect on senior management commitment.

S1-30   Which of the following **MOST** commonly falls within the scope of an information security governance steering committee?

A.   Interviewing candidates for information security specialist positions
B.   Developing content for security awareness programs
C.   Prioritizing information security initiatives
D.   Approving access to critical financial systems

C   Prioritizing information security initiatives is the only appropriate item. The interviewing of specialists should be performed by the information security manager, while the developing of program content should be performed by the information security staff. Approving access to critical financial systems is the responsibility of individual system data owners.

S1-31   Which of the following is the **MOST** important factor when designing information security architecture?

A.   Technical platform interfaces
B.   Scalability of the network
C.   Development methodologies
D.   Stakeholder requirements

D   The most important factor for information security is that it advances the interests of the business, as defined by stakeholder requirements. Interoperability and scalability, as well as development methodologies, are all important but are without merit if a technologically-elegant solution is achieved that does not meet the needs of the business.

S1-32   Who should be responsible for enforcing access rights to application data?

A.   Data owners
B.   Business process owners
C.   The security steering committee
D.   Security administrators

D   As custodians, security administrators are responsible for enforcing access rights to data. Data owners are responsible for approving these access rights. Business process owners are sometimes the data owners as well, and would not be responsible for enforcement. The security steering committee would not be responsible for enforcement.

S1-33   Which of the following is the **MOST** appropriate task for a chief information security officer (CISO) to perform?

A.   Update platform-level security settings
B.   Conduct disaster recovery test exercises
C.   Approve access to critical financial systems
D.   Develop an information security strategy

D   Developing a strategy for information security would be the most appropriate task. Approving access would be the job of the data owner. Updating platform-level security and conducting recovery test exercises would typically be performed by lower-level personnel since these are basic administrative tasks.

S1-34   When an information security manager is developing a strategic plan for information security, the timeline for the plan should be:

A.   aligned with the IT strategic plan.
B.   based on the current rate of technological change.
C.   three-to-five years for both hardware and software.
D.   aligned with the business strategy.

D   Any planning for information security should be properly aligned with the needs of the business. Technology should not come before the needs of the business, nor should planning be done on an artificial timetable that ignores business needs.

S1-35   Which of the following is the **MOST** important information to include in a strategic plan for information security?

A.   Information security staffing requirements
B.   Current state and desired future state
C.   IT capital investment requirements
D.   Information security mission statement

B   It is most important to paint a vision for the future and then draw a road map from the starting point to the desired future state. Staffing, capital investment and the mission all stem from this foundation.

S1-36   Information security projects should be prioritized on the basis of:

A. time required for implementation.
B. impact on the organization.
C. total cost for implementation.
D. mix of resources required.

B   Information security projects should be assessed on the basis of the positive impact that they will have on the organization. Time, cost and resource issues should be subordinate to this objective.

S1-37   Which of the following would **BEST** prepare an information security manager for regulatory reviews?

A. Assign an information security administrator as regulatory liaison
B. Perform self-assessments using regulatory guidelines and reports
C. Assess previous regulatory reports with process owners input
D. Ensure all regulatory inquiries are sanctioned by the legal department

B   Self-assessments provide the best feedback on readiness and permit identification of items requiring remediation. Directing regulators to a specific person or department, or assessing previous reports, is not as effective. The legal department should review all formal inquiries but this does not help prepare for a regulatory review.

S1-38   From an information security manager perspective, what is the immediate benefit of clearly-defined roles and responsibilities?

A. Enhanced policy compliance
B. Improved procedure flows
C. Segregation of duties
D. Better accountability

D   Without well-defined roles and responsibilities, there cannot be accountability. Choice A is incorrect because policy compliance requires adequately defined accountability first and therefore is a byproduct. Choice B is incorrect because people can be assigned to execute procedures that are not well designed. Choice C is incorrect because segregation of duties is not automatic, and roles may still include conflicting duties.

S1-39   Which of the following is responsible for legal and regulatory liability?

A. Chief security officer (CSO)
B. Chief legal counsel (CLC)
C. Board and senior management
D. Information security steering group

C   The board of directors and senior management are ultimately responsible for all that happens in the organization. The others are not individually liable for failures of security in the organization.

S1-40   While implementing information security governance an organization should **FIRST**:

A. adopt security standards.
B. determine security baselines.
C. define the security strategy.
D. establish security policies.

C   The first step in implementing information security governance is to define the security strategy based on which security baselines are determined. Adopting suitable security standards, performing risk assessment and implementing security policy are steps that follow the definition of the security strategy.

S1-41   The **MOST** basic requirement for an information security governance program is to:

A. be aligned with the corporate business strategy.
B. be based on a sound risk management approach.
C. provide adequate regulatory compliance.
D. provide best practices for security initiatives.

A   To receive senior management support, an information security program should be aligned with the corporate business strategy. Risk management is a requirement of an information security program which should take into consideration the business strategy. Security governance is much broader than just regulatory compliance. Best practice is an operational concern and does not have a direct impact on a governance program.

S1-42   Information security policy enforcement is the responsibility of the:

A. security steering committee.
B. chief information officer (CIO).
C. chief information security officer (CISO).
D. chief compliance officer (CCO).

C   Information security policy enforcement is the responsibility of the chief information security officer (CISO), first and foremost. The board of directors and executive management should ensure that a security policy is in line with corporate objectives. The chief information officer (CIO) and the chief compliance officer (CCO) are involved in the enforcement of the policy but are not directly responsible for it.

S1-43   When designing an information security quarterly report to management, the **MOST** important element to be considered should be the:

A. information security metrics.
B. knowledge required to analyze each issue.
C. linkage to business area objectives.
D. baseline against which metrics are evaluated.

C   The link to business objectives is the most important element that would be considered by management. Information security metrics should be put in the context of impact to management objectives. Although important, the security knowledge required would not be the first element to be considered. Baselining against the information security metrics will be considered later in the process.

S1-44   An information security manager at a global organization has to ensure that the local information security program will initially ensure compliance with the:

A.   corporate data privacy policy.
B.   data privacy policy where data are collected.
C.   data privacy policy of the headquarters' country.
D.   data privacy directive applicable globally.

B   As a subsidiary, the local entity will have to comply with the local law for data collected in the country. Senior management will be accountable for this legal compliance. The policy, being internal, cannot supersede the local law. Additionally, with local regulations differing from the country in which the organization is headquartered, it is improbable that a groupwide policy will address all the local legal requirements. In case of data collected locally (and potentially transferred to a country with a different data privacy regulation), the local law applies, not the law applicable to the head office. The data privacy laws are country-specific.

S1-45   A new regulation for safeguarding information processed by a specific type of transaction has come to the attention of an information security officer. The officer should **FIRST**:

A.   meet with stakeholders to decide how to comply.
B.   analyze key risks in the compliance process.
C.   assess whether existing controls meet the regulation.
D.   update the existing security/privacy policy.

C   If the organization is in compliance through existing controls, the need to perform other work related to the regulation is not a priority. The other choices are appropriate and important; however, they are actions that are subsequent and will depend on whether there is an existing control gap.

S1-46   The **PRIMARY** objective of a security steering group is to:

A.   ensure information security covers all business functions.
B.   ensure information security aligns with business goals.
C.   raise information security awareness across the organization.
D.   implement all decisions on security management across the organization.

B   The security steering group comprises senior management of key business functions and has the primary objective to align the security strategy with the business direction. Option A is incorrect because all business areas may not be required to be covered by information security; but, if they do, the main purpose of the steering committee would be alignment more so than coverage. While raising awareness is important, this goal would not be carried out by the committee itself. The steering committee may delegate part of the decision making to the information security manager; however, if it retains this authority, it is not the primary goal.

S1-47   Data owners must provide a safe and secure environment to ensure confidentiality, integrity and availability of the transaction. This is an example of an information security:

   A.   baseline.
   B.   strategy.
   C.   procedure.
   D.   policy.

D   A policy is a high-level statement of an organization's beliefs, goals, roles and objectives. Baselines assume a minimum security level throughout an organization. The information security strategy aligns the information security program with business objectives rather than making control statements. A procedure is a step-by-step process of how policy and standards will be implemented.

S1-48   A security manager is preparing a report to obtain the commitment of executive management to a security program. Inclusion of which of the following would be of **MOST** value?

   A.   Examples of genuine incidents at similar organizations
   B.   Statement of generally accepted best practices
   C.   Associating realistic threats to corporate objectives
   D.   Analysis of current technological exposures

C   Linking realistic threats to key business objectives will direct executive attention to them. All other options are supportive but not of as great a value as choice C when trying to obtain the funds for a new program.

S1-49   The **PRIMARY** concern of an information security manager documenting a formal data retention policy would be:

   A.   generally accepted industry best practices.
   B.   business requirements.
   C.   legislative and regulatory requirements.
   D.   storage availability.

B   The primary concern will be to comply with legislation and regulation but only if this is a genuine business requirement. Best practices may be a useful guide but not a primary concern. Legislative and regulatory requirements are only relevant if compliance is a business need. Storage is irrelevant since whatever is needed must be provided

S1-50   When personal information is transmitted across networks, there MUST be adequate controls over:

   A.   change management.
   B.   privacy protection.
   C.   consent to data transfer.
   D.   encryption devices.

B   Privacy protection is necessary to ensure that the receiving party has the appropriate level of protection of personal data. Change management primarily protects only the information, not the privacy of the individuals. Consent is one of the protections that is frequently, but not always, required. Encryption is a method of achieving the actual control, but controls over the devices may not ensure adequate privacy protection and, therefore, is a partial answer.

S1-51   Who in an organization has the responsibility for classifying information?

      A. Data custodian
      B. Database administrator
      C. Information security officer
      D. Data owner

D   The data owner has full responsibility over data. The data custodian is responsible for securing the information. The database administrator carries out the technical administration. The information security officer oversees the overall classification management of the information.

S1-52   What is the **PRIMARY** role of the information security manager in the process of information classification within an organization?

      A. Defining and ratifying the classification structure of information assets
      B. Deciding the classification levels applied to the organization's information assets
      C. Securing information assets in accordance with their classification
      D. Checking if information assets have been classified properly

A   Defining and ratifying the classification structure of information assets is the primary role of the information security manager in the process of information classification within the organization. Choice B is incorrect because the final responsibility for deciding the classification levels rests with the data owners. Choice C is incorrect because the job of securing information assets is the responsibility of the data custodians. Choice D may be a role of an information security manager but is not the key role in this context.

S1-53   Which of the following is **MOST** important in developing a security strategy?

      A. Creating a positive business security environment
      B. Understanding key business objectives
      C. Having a reporting line to senior management
      D. Allocating sufficient resources to information security

B   Alignment with business strategy is of utmost importance. Understanding business objectives is critical in determining the security needs of the organization.

S1-54   Who is ultimately responsible for the organization's information?

      A. Data custodian
      B. Chief information security officer (CISO)
      C. Board of directors
      D. Chief information officer (CIO)

C   The board of directors is ultimately responsible for the organization's information and is tasked with responding to issues that affect its protection. The data custodian is responsible for the maintenance and protection of data. This role is usually filled by the IT department. The chief information security officer (CISO) is responsible for security and carrying out senior management's directives. The chief information officer (CIO) is responsible for information technology within the organization and is not ultimately responsible for the organization's information.

S1-55   An information security manager mapping a job description to types of data access is **MOST** likely to adhere to which of the following information security principles?

    A. Ethics
    B. Proportionality
    C. Integration
    D. Accountability

B   Information security controls should be proportionate to the risks of modification, denial of use or disclosure of the information. It is advisable to learn if the job description is apportioning more data than are necessary for that position to execute the business rules (types of data access). Principles of ethics and integration have the least to do with mapping job description to types of data access. The principle of accountability would be the second most adhered to principle since people with access to data may not always be accountable but may be required to perform an operation.

S1-56   Which of the following is the **MOST** important prerequisite for establishing information security management within an organization?

    A. Senior management commitment
    B. Information security framework
    C. Information security organizational structure
    D. Information security policy

A   Senior management commitment is necessary in order for each of the other elements to succeed. Without senior management commitment, the other elements will likely be ignored within the organization.

S1-57   What will have the **HIGHEST** impact on standard information security governance models?

    A. Number of employees
    B. Distance between physical locations
    C. Complexity of organizational structure
    D. Organizational budget

C   Information security governance models are highly dependent on the overall organizational structure. Some of the elements that impact organizational structure are multiple missions and functions across the organization, leadership and lines of communication. Number of employees and distance between physical locations have less impact on information security governance models since well-defined process, technology and people components intermingle to provide the proper governance. Organizational budget is not a major impact once good governance models are in place, hence governance will help in effective management of the organization's budget.

S1-58   In order to highlight to management the importance of integrating information security in the business processes, a newly hired information security officer should **FIRST**:

    A. prepare a security budget.
    B. conduct a risk assessment.
    C. develop an information security policy.
    D. obtain benchmarking information.

B   Risk assessment, evaluation and impact analysis will be the starting point for driving management's attention to information security. All other choices will follow the risk assessment.

S1-59   An outcome of effective security governance is:

   A. business dependency assessment.
   B. strategic alignment.
   C. risk assessment.
   D. planning.

B   Business dependency assessment is a process of determining the dependency of a business on certain
              information resources. It is not an outcome or a product of effective security management. Strategic alignment
              is an outcome of effective security governance. Where there is good governance, there is likely to be strategic
              alignment. Risk assessment is not an outcome of effective security governance; it is a process. Planning comes
              at the beginning of effective security governance, and is not an outcome but a process.

S1-60   How would an information security manager balance the potentially conflicting requirements of an
              international organization's security standards and local regulation?

   A. Give organization standards preference over local regulations
   B. Follow local regulations only
   C. Make the organization aware of those standards where local regulations causes conflicts
   D. Negotiate a local version of the organization standards

D   Adherence to local regulations must always be the priority. Not following local regulations can prove
              detrimental to the group organization. Following local regulations only is incorrect since there needs to be
              some recognition of organization requirements. Making an organization aware of standards is a sensible
              step, but is not a total solution. Negotiating a local version of the organization standards is the most
              effective compromise in this situation.

S1-61   Which of the following should drive the risk analysis for an organization?

   A. Senior management
   B. Security manager
   C. Quality manager
   D. Legal department

B   Although senior management should support and sponsor a risk analysis, the know-how and the
              management of the project will be with the security department. Quality management and the legal
              department will contribute to the project.

S1-62   The **FIRST** step in developing an information security management program is to:

   A. identify business risks that affect the organization.
   B. clarify organizational purpose for creating the program.
   C. assign responsibility for the program.
   D. assess adequacy of controls to mitigate business risks.

B   In developing an information security management program, the first step is to clarify the organization's
              purpose for creating the program. This is a business decision based more on judgment than on any specific
              quantitative measures. After clarifying the purpose, the other choices are assigned and acted upon.

S1-63   What would a security manager **PRIMARILY** utilize when proposing the implementation of a
         security solution?

         A. Risk assessment report
         B. Technical evaluation report
         C. Business case
         D. Budgetary requirements

C   The information security manager needs to prioritize the controls based on risk management and the
         requirements of the organization. The information security manager must look at the costs of the various
         controls and compare them against the benefit the organization will receive from the security solution. The
         information security manager needs to have knowledge of the development of business cases to illustrate
         the costs and benefits of the various controls. All other choices are supplemental.

S1-64   To justify its ongoing security budget, which of the following would be of **MOST** use to the information
         security department?

         A. Security breach frequency
         B. Annualized loss expectancy (ALE)
         C. Cost-benefit analysis
         D. Peer group comparison

C   Cost-benefit analysis is the legitimate way to justify budget. The frequency of security breaches may assist
         the argument for budget but is not the key tool; it does not address the impact. Annualized loss expectancy
         (ALE) does not address the potential benefit of security investment. Peer group comparison would provide
         a good estimate for the necessary security budget but it would not take into account the specific needs of
         the organization.

S1-65   Which of the following situations would **MOST** inhibit the effective implementation of security governance:

         A. The complexity of technology
         B. Budgetary constraints
         C. Conflicting business priorities
         D. High-level sponsorship

D   The need for senior management involvement and support is a key success factor for the implementation of
         appropriate security governance. Complexity of technology, budgetary constraints and conflicting business
         priorities are realities that should be factored into the governance model of the organization, and should not
         be regarded as inhibitors.

S1-66   To achieve effective strategic alignment of security initiatives, it is important that:

         A. steering committee leadership be selected by rotation.
         B. inputs be obtained and consensus achieved between the major organizational units.
         C. the business strategy be updated periodically.
         D. procedures and standards be approved by all departmental heads.

B   It is important to achieve consensus on risks and controls, and obtain inputs from various organizational
         entities since security needs to be aligned to the needs of the organization. Rotation of steering committee
         leadership does not help in achieving strategic alignment. Updating business strategy does not lead
         to strategic alignment of security initiatives. Procedures and standards need not be approved by all
         departmental heads

S1-67   In implementing information security governance, the information security manager is **PRIMARILY** responsible for:

A. developing the security strategy.
B. reviewing the security strategy.
C. communicating the security strategy.
D. approving the security strategy

A   The information security manager is responsible for developing a security strategy based on business objectives with the help of business process owners. Reviewing the security strategy is the responsibility of a steering committee. The information security manager is not necessarily responsible for communicating or approving the security strategy.

S1-68   An information security strategy document that includes specific links to an organization's business activities is **PRIMARILY** an indicator of:

A. performance measurement.
B. integration.
C. alignment.
D. value delivery.

C   Strategic alignment of security with business objectives is a key indicator of performance measurement. In guiding a security program, a meaningful performance measurement will also rely on an understanding of business objectives, which will be an outcome of alignment. Business linkages do not by themselves indicate integration or value delivery. While alignment is an important precondition, it is not as important an indicator.

S1-69   To justify the need to invest in a forensic analysis tool, an information security manager should **FIRST**:

A. review the functionalities and implementation requirements of the solution.
B. review comparison reports of tool implementation in peer companies.
C. provide examples of situations where such a tool would be useful.
D. demonstrate that the investment meets organizational needs.

D   Any investment must be reviewed to determine whether it is cost effective and supports the organizational strategy. It is important to review the features and functionalities provided by such a tool, and to provide examples of situations where the tool would be useful, but that comes after substantiating the investment and return on investment to the organization.

S1-70   The **MOST** useful way to describe the objectives in the information security strategy is through:

A. attributes and characteristics of the 'desired state.'
B. overall control objectives of the security program.
C. mapping the IT systems to key business processes.
D. calculation of annual loss expectations.

A   Security strategy will typically cover a wide variety of issues, processes, technologies and outcomes that can best be described by a set of characteristics and attributes that are desired. Control objectives are developed after strategy and policy development. Mapping IT systems to key business processes does not address strategy issues. Calculation of annual loss expectations would not describe the objectives in the information security strategy.

S1-71   In order to highlight to management the importance of network security, the security manager should **FIRST**:

A. develop a security architecture.
B. install a network intrusion detection system (NIDS) and prepare a list of attacks.
C. develop a network security policy.
D. conduct a risk assessment.

D   A risk assessment would be most helpful to management in understanding at a very high level the threats, probabilities and existing controls. Developing a security architecture, installing a network intrusion detection system (NIDS) and preparing a list of attacks on the network and developing a network security policy would not be as effective in highlighting the importance to management and would follow only after performing a risk assessment.

S1-72   The **MOST** important characteristic of good security policies is that they:

A. state expectations of IT management.
B. state only one general security mandate.
C. are aligned with organizational goals.
D. govern the creation of procedures and guidelines.

C   The most important characteristic of good security policies is that they be aligned with organizational goals. Failure to align policies and goals significantly reduces the value provided by the policies. Stating expectations of IT management omits addressing overall organizational goals and objectives. Stating only one general security mandate is the next best option since policies should be clear; otherwise, policies may be confusing and difficult to understand. Governing the creation of procedures and guidelines is most relevant to information security standards.

S1-73   An information security manager must understand the relationship between information security and business operations in order to:

A. support organizational objectives.
B. determine likely areas of noncompliance.
C. assess the possible impacts of compromise.
D. understand the threats to the business.

A   Security exists to provide a level of predictability for operations, support for the activities of the organization and to ensure preservation of the organization. Business operations must be the driver for security activities in order to set meaningful objectives, determine and manage the risks to those activities, and provide a basis to measure the effectiveness of and provide guidance to the security program. Regulatory compliance may or may not be an organizational requirement. If compliance is a requirement, some level of compliance must be supported but compliance is only one aspect. It is necessary to understand the business goals in order to assess potential impacts and evaluate threats. These are some of the ways in which security supports organizational objectives, but they are not the only ways.

S1-74   The **MOST** effective approach to address issues that arise between IT management, business units and security management when implementing a new security strategy is for the information security manager to:

        A.   escalate issues to an external third party for resolution.
        B.   ensure that senior management provide authority for security to address the issues.
        C.   insist that managers or units not in agreement with the security solution accept the risk.
        D.   refer the issues to senior management along with any security recommendations.

D   Senior management is in the best position to arbitrate since they will look at the overall needs of the business in reaching a decision. The authority may be delegated to others by senior management after their review of the issues and security recommendations. Units should not be asked to accept the risk without first receiving input from senior management.

S1-75   Obtaining senior management support for establishing a warm site can **BEST** be accomplished by:

        A.   establishing a periodic risk assessment.
        B.   promoting regulatory requirements.
        C.   developing a business case.
        D.   developing effective metrics.

C   Business case development, including a cost-benefit analysis, will be most persuasive to management. A risk assessment may be included in the business case, but by itself will not be as effective in gaining management support. Informing management of regulatory requirements may help gain support for initiatives, but given that more than half of all organizations are not in compliance with regulations, it is unlikely to be sufficient in many cases. Good metrics which provide assurance that initiatives are meeting organizational goals will also be useful, but are insufficient in gaining management support.

S1-76   Which of the following would be the **BEST** option to improve accountability for a system administrator who has security functions?

        A.   Include security responsibilities in the job description
        B.   Require the administrator to obtain security certification
        C.   Train the system administrator on penetration testing and vulnerability assessment
        D.   Train the system administrator on risk assessment

A   The first step to improve accountability is to include security responsibilities in a job description. This documents what is expected and approved by the organization. The other choices are methods to ensure that the system administrator has the training to fulfill the responsibilities included in the job description.

S1-77   Which of the following is the **MOST** important element of an information security strategy?

        A.   Defined objectives
        B.   Time frames for delivery
        C.   Adoption of a control framework
        D.   Complete policies

A   Without defined objectives, a strategy—the plan to achieve objectives—cannot be developed. Time frames for delivery are important but not critical for inclusion in the strategy document. Similarly, the adoption of a control framework is not critical to having a successful information security strategy. Policies are developed subsequent to, and as a part of, implementing a strategy.

S1-78 A multinational organization operating in fifteen countries is considering implementing an information security program. Which factor will **MOST** influence the design of the Information security program?

A. Representation by regional business leaders
B. Composition of the board
C. Cultures of the different countries
D. IT security skills

C Culture has a significant impact on how information security will be implemented. Representation by regional business leaders may not have a major influence unless it concerns cultural issues. Composition of the board may not have a significant impact compared to cultural issues. IT security skills are not as key or high impact in designing a multinational information security program as would be cultural issues.

S1-79 Which of the following is the **BEST** justification to convince management to invest in an information security program?

A. Cost reduction
B. Compliance with company policies
C. Protection of business assets
D. Increased business value

D Investing in an information security program should increase business value and confidence. Cost reduction by itself is rarely the motivator for implementing an information security program. Compliance is secondary to business value. Increasing business value may include protection of business assets.

S1-80 On a company's e-commerce web site, a good legal statement regarding data privacy should include:

A. a statement regarding what the company will do with the information it collects.
B. a disclaimer regarding the accuracy of information on its web site.
C. technical information regarding how information is protected.
D. a statement regarding where the information is being hosted.

A Most privacy laws and regulations require disclosure on how information will be used. A disclaimer is not necessary since it does not refer to data privacy. Technical details regarding how information is protected are not mandatory to publish on the web site and in fact would not be desirable. It is not mandatory to say where information is being hosted.

S1-81 The **MOST** important factor in ensuring the success of an information security program is effective:

A. communication of information security requirements to all users in the organization.
B. formulation of policies and procedures for information security.
C. alignment with organizational goals and objectives .
D. monitoring compliance with information security policies and procedures.

C The success of security programs is dependent upon alignment with organizational goals and objectives. Communication is a secondary step. Effective communication and education of users is a critical determinant of success but alignment with organizational goals and objectives is the most important factor for success. Mere formulation of policies without effective communication to users will not ensure success. Monitoring compliance with information security policies and procedures can be, at best, a detective mechanism that will not lead to success in the midst of uninformed users.

S1-82   Which of the following would be **MOST** helpful to achieve alignment between information security and organization objectives?

A. Key control monitoring
B. A robust security awareness program
C. A security program that enables business activities
D. An effective security architecture

C   A security program enabling business activities would be most helpful to achieve alignment between information security and organization objectives. All of the other choices are part of the security program and would not individually and directly help as much as the security program.

S1-83   Which of the following **BEST** contributes to the development of a security governance framework that supports the maturity model concept?

A. Continuous analysis, monitoring and feedback
B. Continuous monitoring of the return on security investment (ROSI)
C. Continuous risk reduction
D. Key risk indicator (KRI) setup to security management processes

A   To improve the governance framework and achieve a higher level of maturity, an organization needs to conduct continuous analysis, monitoring and feedback compared to the current state of maturity. Return on security investment (ROSI) may show the performance result of the security-related activities; however, the result is interpreted in terms of money and extends to multiple facets of security initiatives. Thus, it may not be an adequate option. Continuous risk reduction would demonstrate the effectiveness of the security governance framework, but does not indicate a higher level of maturity. Key risk indicator (KRI) setup is a tool to be used in internal control assessment. KRI setup presents a threshold to alert management when controls are being compromised in business processes. This is a control tool rather than a maturity model support tool.

S1-84   The **MOST** complete business case for security solutions is one that:

A. includes appropriate justification.
B. explains the current risk profile.
C. details regulatory requirements.
D. identifies incidents and losses.

A   Management is primarily interested in security solutions that can address risks in the most cost-effective way. To address the needs of an organization, a business case should address appropriate security solutions in line with the organizational strategy.

S1-85   Which of the following is **MOST** important to understand when developing a meaningful information security strategy?

A. Regulatory environment
B. International security standards
C. Organizational risks
D. Organizational goals

D   Alignment of security with business objectives requires an understanding of what an organization is trying to accomplish. The other choices are all elements that must be considered, but their importance is secondary and will vary depending on organizational goals.

S1-86   Which of the following would help to change an organization's security culture?

A. Develop procedures to enforce the information security policy
B. Obtain strong management support
C. Implement strict technical security controls
D. Periodically audit compliance with the information security policy

B   Management support and pressure will help to change an organization's culture. Procedures will support an information security policy, but cannot change the culture of the organization. Technical controls will provide more security to an information system and staff; however, this does not mean the culture will be changed. Auditing will help to ensure the effectiveness of the information security policy; however, auditing is not effective in changing the culture of the company.

S1-87   The **BEST** way to justify the implementation of a single sign-on (SSO) product is to use:

A. return on investment (ROI).
B. a vulnerability assessment.
C. annual loss expectancy (ALE).
D. a business case.

D   A business case shows both direct and indirect benefits, along with the investment required and the expected returns, thus making it useful to present to senior management. Return on investment (ROI) would only provide the costs needed to preclude specific risks, and would not provide other indirect benefits such as process improvement and learning. A vulnerability assessment is more technical in nature and would only identify and assess the vulnerabilities. This would also not provide insights on indirect benefits. Annual loss expectancy (ALE) would not weigh the advantages of implementing single sign-on (SSO) in comparison to the cost of implementation.

S1-88   The **FIRST** step in establishing a security governance program is to:

A. conduct a risk assessment.
B. conduct a workshop for all end users.
C. prepare a security budget.
D. obtain high-level sponsorship.

D   The establishment of a security governance program is possible only with the support and sponsorship of top management since security governance projects are enterprisewide and integrated into business processes. Conducting a risk assessment, conducting a workshop for all end users and preparing a security budget all follow once high-level sponsorship is obtained.

S1-89   An IS manager has decided to implement a security system to monitor access to the Internet and prevent access to numerous sites. Immediately upon installation, employees flood the IT helpdesk with complaints of being unable to perform business functions on Internet sites. This is an example of:

A. conflicting security controls with organizational needs.
B. strong protection of information resources.
C. implementing appropriate controls to reduce risk.
D. proving information security's protective abilities.

A   The needs of the organization were not taken into account, so there is a conflict. This example is not strong protection, it is poorly configured. Implementing appropriate controls to reduce risk is not an appropriate control as it is being used. This does not prove the ability to protect, but proves the ability to interfere with business.

S1-90   An organization's information security strategy should be based on:

    A. managing risk relative to business objectives.
    B. managing risk to a zero level and minimizing insurance premiums.
    C. avoiding occurrence of risks so that insurance is not required.
    D. transferring most risks to insurers and saving on control costs.

A   Organizations must manage risks to a level that is acceptable for their business model, goals and objectives. A zero-level approach may be costly and not provide the effective benefit of additional revenue to the organization. Long-term maintenance of this approach may not be cost effective. Risks vary as business models, geography, and regulatory and operational processes change. Insurance covers only a small portion of risks and requires that the organization have certain operational controls in place.

S1-91   Which of the following should be included in an annual information security budget that is submitted for management approval?

    A. A cost-benefit analysis of budgeted resources
    B. All of the resources that are recommended by the business
    C. Total cost of ownership (TCO)
    D. Baseline comparisons

A   A brief explanation of the benefit of expenditures in the budget helps to convey the context of how the purchases that are being requested meet goals and objectives, which in turn helps build credibility for the information security function or program. Explanations of benefits also help engage senior management in the support of the information security program. While the budget should consider all inputs and recommendations that are received from the business, the budget that is ultimately submitted to management for approval should include only those elements that are intended for purchase. TCO may be requested by management and may be provided in an addendum to a given purchase request, but is not usually included in an annual budget. Baseline comparisons (cost comparisons with other companies or industries) may be useful in developing a budget or providing justification in an internal review for an individual purchase, but would not be included with a request for budget approval.

S1-92   Which of the following is a benefit of information security governance?

    A. Reduction of the potential for civil or legal liability
    B. Questioning trust in vendor relationships
    C. Increasing the risk of decisions based on incomplete management information
    D. Direct involvement of senior management in developing control processes

A   Information security governance decreases the risk of civil or legal liability. The remaining answers are incorrect. Option D appears to be correct, but senior management would provide oversight and approval as opposed to direct involvement in developing control processes.

S1-93   Investment in security technology and processes should be based on:

     A. clear alignment with the goals and objectives of the organization.
     B. success cases that have been experienced in previous projects.
     C. best business practices.
     D. safeguards that are inherent in existing technology.

A   Organization maturity level for the protection of information is a clear alignment with goals and objectives of the organization. Experience in previous projects is dependent upon other business models which may not be applicable to the current model. Best business practices may not be applicable to the organization's business needs. Safeguards inherent to existing technology are low cost but may not address all business needs and/or goals of the organization.

S1-94   The data access requirements for an application should be determined by the:

     A. legal department.
     B. compliance officer.
     C. information security manager.
     D. business owner.

D   Business owners are ultimately responsible for their applications. The legal department, compliance officer and information security manager all can advise, but do not have final responsibility.

S1-95   From an information security perspective, information that no longer supports the main purpose of the business should be:

     A. analyzed under the retention policy.
     B. protected under the information classification policy.
     C. analyzed under the backup policy.
     D. assessed by a business impact analysis (BIA).

A   Choice A is the type of analysis that will determine whether the organization is required to maintain the data for business, legal or regulatory reasons. Keeping data that are no longer required unnecessarily consumes resources, and, in the case of sensitive personal information, can increase the risk of data compromise. Choices B, C and D are attributes that should be considered in the destruction and retention policy. A BIA could help determine that this information does not support the main objective of the business, but does not indicate the action to take.

S1-96   Effective IT governance is **BEST** ensured by:

    A. utilizing a bottom-up approach.
    B. management by the IT department.
    C. referring the matter to the organization's legal department.
    D. utilizing a top-down approach.

D   Effective IT governance needs to be a top-down initiative, with the board and executive management setting clear policies, goals and objectives and providing for ongoing monitoring of the same. Focus on the regulatory issues and management priorities may not be reflected effectively by a bottom-up approach. IT governance affects the entire organization and is not a matter concerning only the management of IT. The legal department is part of the overall governance process, but cannot take full responsibility.

S1-97   The **FIRST** step to create an internal culture that focuses on information security is to:

    A. implement stronger controls.
    B. conduct periodic awareness training.
    C. actively monitor operations.
    D. gain the endorsement of executive management.

D   Endorsement of executive management in the form of policies provides direction and awareness. The implementation of stronger controls may lead to circumvention. Awareness training is important, but must be based on policies. Actively monitoring operations will not affect culture at all levels.

S1-98   Which of the following is the **BEST** method or technique to ensure the effective implementation of an information security program?

    A. Obtain the support of the board of directors.
    B. Improve the content of the information security awareness program.
    C. Improve the employees' knowledge of security policies.
    D. Implement logical access controls to the information systems.

A   It is extremely difficult to implement an information security program without the aid and support of the board of directors. If they do not understand the importance of security to the achievement of the business objectives, other measures will not be sufficient. Choices B and C are measures proposed to ensure the efficiency of the information security program implementation, but are of less significance than obtaining the aid and support of the board of directors. Choice D is a measure to secure the enterprise information, but by itself is not a measure to ensure the broader effectiveness of an information security program.

S1-99   When an organization is implementing an information security governance program, its board of directors should be responsible for:

    A. drafting information security policies.
    B. reviewing training and awareness programs.
    C. setting the strategic direction of the program.
    D. auditing for compliance.

C   A board of directors should establish the strategic direction of the program to ensure that it is in sync with the company's vision and business goals. The board must incorporate the governance program into the overall corporate business strategy. Drafting information security policies is best fulfilled by someone such as a security manager with the expertise to bring balance, scope and focus to the policies. Reviewing training and awareness programs may best be handled by security management and training staff to ensure that the training is on point and follows best practices. Auditing for compliance is best left to the internal and external auditors to provide an objective review of the program and how it meets regulatory and statutory compliance.

S1-100   A risk assessment and business impact analysis (BIA) have been completed for a major proposed purchase and new process for an organization. There is disagreement between the information security manager and the business department manager who will own the process regarding the results and the assigned risk. Which of the following would be the **BEST** approach of the information security manager?

    A. Acceptance of the business manager's decision on the risk to the corporation
    B. Acceptance of the information security manager's decision on the risk to the corporation
    C. Review of the assessment with executive management for final input
    D. A new risk assessment and BIA are needed to resolve the disagreement

C   Executive management must be supportive of the process and fully understand and agree with the results since risk management decisions can often have a large financial impact and require major changes. Risk management means different things to different people, depending upon their role in the organization, so the input of executive management is important to the process.

S1-101   Who is responsible for ensuring that information is categorized and that specific protective measures are taken?

    A. The security officer
    B. Senior management
    C. The end user
    D. The custodian

B   Routine administration of all aspects of security is delegated, but top management must retain overall responsibility. The security officer supports and implements information security for senior management. The end user does not perform categorization. The custodian supports and implements information security measures as directed.

S1-102   An organization's board of directors has learned of recent legislation requiring organizations within the industry to enact specific safeguards to protect confidential customer information. What actions should the board take next?

A. Direct information security on what they need to do
B. Research solutions to determine the proper solutions
C. Require management to report on compliance
D. Nothing; information security does not report to the board

C   Information security governance is the responsibility of the board of directors and executive management. In this instance, the appropriate action is to ensure that a plan is in place for implementation of needed safeguards and to require updates on that implementation.

S1-103   Information security should be:

A. focused on eliminating all risks.
B. a balance between technical and business requirements.
C. driven by regulatory requirements.
D. defined by the board of directors.

B   Information security should ensure that business objectives are met given available technical capabilities, resource constraints and compliance requirements. It is not practical or feasible to eliminate all risks. Regulatory requirements must be considered, but are inputs to the business considerations. The board of directors does not define information security, but provides direction in support of the business goals and objectives.

S1-104   What is the **MOST** important factor in the successful implementation of an enterprisewide information security program?

A. Realistic budget estimates
B. Security awareness
C. Support of senior management
D. Recalculation of the work factor

C   Without the support of senior management, an information security program has little chance of survival. A company's leadership group, more than any other group, will more successfully drive the program. Their authoritative position in the company is a key factor. Budget approval, resource commitments, and companywide participation also require the buy-in from senior management. Senior management is responsible for providing an adequate budget and the necessary resources. Security awareness is important, but not the most important factor. Recalculation of the work factor is a part of risk management.

S1-105   What is the **MAIN** risk when there is no user management representation on the Information Security Steering Committee?

A. Functional requirements are not adequately considered.
B. User training programs may be inadequate.
C. Budgets allocated to business units are not appropriate.
D. Information security plans are not aligned with business requirements

D   The steering committee controls the execution of the information security strategy, according to the needs of the organization, and decides on the project prioritization and the execution plan. User management is an important group that should be represented to ensure that the information security plans are aligned with the business needs. Functional requirements and user training programs are considered to be part of the projects but are not the main risks. The steering committee does not approve budgets for business units.

S1-106   The **MAIN** reason for having the Information Security Steering Committee review a new security controls implementation plan is to ensure that:

A. the plan aligns with the organization's business plan.
B. departmental budgets are allocated appropriately to pay for the plan.
C. regulatory oversight requirements are met.
D. the impact of the plan on the business units is reduced.

A   The steering committee controls the execution of the information security strategy according to the needs of the organization and decides on the project prioritization and the execution plan. The steering committee does not allocate department budgets for business units. While ensuring that regulatory oversight requirements are met could be a consideration, it is not the main reason for the review. Reducing the impact on the business units is a secondary concern but not the main reason for the review.

S1-107   Which of the following should be determined while defining risk management strategies?

A. Risk assessment criteria
B. Organizational objectives and risk appetite
C. IT architecture complexity
D. Enterprise disaster recovery plans

B   While defining risk management strategies, one needs to analyze the organization's objectives and risk appetite and define a risk management framework based on this analysis. Some organizations may accept known risks, while others may invest in and apply mitigation controls to reduce risks. Risk assessment criteria would become part of this framework, but only after proper analysis. IT architecture complexity and enterprise disaster recovery plans are more directly related to assessing risks than defining strategies.

S1-108   When implementing effective security governance within the requirements of the company's security strategy, which of the following is the **MOST** important factor to consider?

A. Preserving the confidentiality of sensitive data
B. Establishing international security standards for data sharing
C. Adhering to corporate privacy standards
D. Establishing system manager responsibility for information security

A   The goal of information security is to protect the organization's information assets. International security standards are situational, depending upon the company and its business. Adhering to corporate privacy standards is important, but those standards must be appropriate and adequate and are not the most important factor to consider. All employees are responsible for information security, but it is not the most important factor to consider.

S1-109   Which of the following is the **BEST** reason to perform a business impact analysis (BIA)?

A. To help determine the current state of risk
B. To budget appropriately for needed controls
C. To satisfy regulatory requirements
D. To analyze the effect on the business

A   The BIA is included as part of the process to determine the current state of risk and helps determine the acceptable levels of response from impacts and the current level of response, leading to a gap analysis. Budgeting appropriately may come as a result, but is not the reason to perform the analysis. Performing an analysis may satisfy regulatory requirements, but is not the reason to perform one. Analyzing the effect on the business is part of the process, but one must also determine the needs or acceptable effect or response.

S1-110   Priority should be given to which of the following to ensure effective implementation of information
          security governance?

          A. Consultation
          B. Negotiation
          C. Facilitation
          D. Planning

D   Planning is the key to effective implementation of information security governance. Consultation,
          negotiation and facilitation come after planning.

S1-111   The director of auditing has recommended a specific information security monitoring solution to the
          information security manager. What should the information security manager do **FIRST**?

          A. Obtain comparative pricing bids and complete the transaction with the vendor offering the best deal.
          B. Add the purchase to the budget during the next budget preparation cycle to account for costs.
          C. Perform an assessment to determine correlation with business goals and objectives.
          D. Form a project team to plan the implementation.

C   An assessment must be made first to determine that the proposed solution is aligned with business goals
          and objectives. The other choices are not necessary until a determination has been made regarding whether
          the product fits the goals and objectives of the business.

S1-112   Of the following, which is the **MOST** effective way to measure strategic alignment of an information
          security program?

          A. Track audits over time.
          B. Evaluate incident losses.
          C. Analyze business cases.
          D. Interview business owners.

D   It is essential that business owners understand and support the security program and fully understand how
          its controls impact their activities. This can be most readily accomplished through direct interaction with
          business leadership. Audit reports may indicate areas of security activities that do not optimally support the
          enterprise objectives, but will not be as good an indicator. Losses may or may not be considered acceptable
          by the enterprise, but will not be well correlated with the perception of business support. To the extent that
          business cases have been developed for particular security activities, they will be a good indication of how
          well business requirements were considered; however, the perception of business owners will ultimately be
          the most important factor.

S1-113   Business objectives should be evident in the security strategy by the presence of:

          A. inferred connections.
          B. traceable connections.
          C. standardized controls.
          D. documented constraints.

B   The security strategy will be most useful if there is a direct traceable connection with business objectives.
          Inferred connections to business objectives are not as good as explicit traceable connections. Standard
          controls may or may not be relevant to a particular business objective. Documenting constraints alone is not
          as useful as also defining explicit benefits.

S1-114 Which of the following is the **MOST** important reason for aligning information security governance with corporate governance?

    A. To show that the information security manager understands the rules
    B. To maximize the cost-effectiveness of controls
    C. To provide operational consistency
    D. To minimize the number of rules and regulations required

B Corporate governance is a structure and corresponding rules that, in most cases, are related to managing various types of risk. A lack of alignment will result in potentially duplicate or contradictory procedural controls, which negatively impacts cost-effectiveness. While it is important that the information security manager understands the corporate governance rules, it is not the main reason for alignment. Operational consistency is just one element of achieving cost-effectiveness. Minimizing the number of rules is helpful, but it is just one element of achieving cost-effectiveness.

S1-115 Which of the following would be the **BEST** approach to securing approval for information security expenditures?

    A. Developing a business case
    B. Conducting a cost-benefit analysis
    C. Calculating return on investment (ROI)
    D. Evaluating loss history

A Justifying and obtaining approval for a security initiative is more likely to be successful if it is supported by a well-developed business case. Choices B, C and D will each be just one typical element of a business case.

S1-116 The formal declaration of organizational security goals and objectives should be found in which of the following documents?

    A. Information security procedures
    B. Information security principles
    C. An employee code of conduct
    D. An information security policy

D An information security policy is management's formal declaration of security goals and objectives. Security procedures are usually detailed as step-by-step actions to ensure that activities meet a given standard. Security principles are not always organizationally specific. A code of conduct is a standard requirement that encompasses more than security.

S1-117 Which of the following would be the **FIRST** step when developing a business case for an information security investment?

    A. Defining the objectives
    B. Calculating the cost
    C. Defining the need
    D. Analyzing the cost-effectiveness

C Without a clear definition of the needs to be fulfilled, the rest of the components of the business case cannot be determined.

S1-118   Laws and regulations should be addressed by the information security manager:

A. to the extent that they impact the enterprise.
B. by implementing international standards.
C. by developing policies that address the requirements.
D. to ensure that guidelines meet the requirements.

A   Legal and regulatory requirements should be assessed based on: the impact of non-compliance or partial compliance balanced against the costs of compliance, the risk tolerance defined by management, and the extent and nature of enforcement. International standards may not address the legal requirements in question. Policies should not address particular regulations because regulations are subject to change. Policies should only address the need to assess regulatory requirements and deal with them appropriately based on risk and impact. Guidelines would normally not address regulations, although standards may address regulations based on management's determination of the appropriate level of compliance.

S1-119   Which of the following is an indicator of effective governance?

A. A defined information security architecture
B. Compliance with international security standards
C. Periodic external audits
D. An established risk management program

D   A risk management program is a key component of effective governance. A defined information security architecture is helpful, but is just an aspect of the risk management program. Compliance with International standards is not an indication of the use of effective governance. Periodic external audits may provide an opinion on effective controls governance.

S1-120   Which of the following **BEST** indicates senior management commitment toward supporting information security?

A. Assessment of various risks to the assets
B. Approval of risk management methodology
C. Review of inherent risks to information assets
D. Review of residual risks for information assets

B   Management sign-off on risk management methodology helps in performing the entire risk cycle. An assessment of various risks to the assets can be done by the security manager and other managers from various departments. A review of inherent risks to information assets can be performed by individual department owners by taking support from IT or from information security. A review of residual risks for information assets is not possible for senior management because there can be lot of information assets, but only business-critical assets can be reviewed.

S1-121   Which of the following is the **MOST** effective approach to identify events that may affect information security across a large multinational enterprise?

A. Review internal and external audits to indicate anomalies.
B. Ensure that intrusion detection sensors are widely deployed.
C. Develop communication channels throughout the enterprise.
D. Conduct regular enterprisewide security reviews.

C   Developing communication channels throughout the enterprise creates a consistent message regarding information security and helps ensure feedback from each business unit. The other choices are tasks that information security managers use to determine or ensure that their programs are effective, but a lack of communication could undermine everything else.

S1-122   Data owners are **PRIMARILY** responsible for:

A. providing access to systems.
B. approving access to systems.
C. establishing authorization and authentication.
D. handling identity management.

B   Approving access to systems is the only answer that fits since choices A and C are the work of data custodians and choice D is the work of the information security staff.

S1-123   The **PRIMARY** objective for information security program development should be:

A. establishing strategic alignment with the business.
B. establishing incident response programs.
C. identifying and implementing the best security solutions.
D. reducing the impact of the risk in the business.

D   Reducing risk to the business is the most important objective of an information security program. Strategic alignment with the business is a characteristic of a sound information security program. The other choices are activities of the security program.

S1-124   During a stakeholder meeting, a question was asked regarding who is ultimately accountable for the protection and security of sensitive data. Assuming all of the choices below exist in the enterprise, which would be the **MOST** appropriate?

A. Security administrators
B. The IT steering committee
C. The board of directors
D. The information security manager

C   The board of directors is ultimately accountable for information security. All other choices have a level of responsibility, but are not ultimately accountable.

S1-125   Which of the following is the **MOST** important objective of an information security strategy review?

A. Ensuring that risks are identified and mitigated
B. Ensuring that information security strategy is aligned with organizational goals
C. Maximizing the return of information security investments
D. Ensuring the efficient utilization of information security resources

B   The most important part of an information security strategy is that it supports the business objectives and goals of the enterprise. The other choices are aspects of the information security strategy.

S1-126   Which of the following is the **BEST** approach to obtain senior management commitment to the information security program?

     A.   Describe the reduction of risk.
     B.   Present the emerging threat environment.
     C.   Benchmark against other enterprises.
     D.   Demonstrate the alignment of the program to business objectives.

D   A security program must be aligned to business objectives. Senior management will support the security program only when it helps achieve the business objectives. The security program will always try to reduce the risk; however, it has to be balanced against the cost and impact to the business. Reduction of risk alone cannot justify the security program from the senior management perspective. The security program monitors emerging threats; however, the threat environment itself cannot determine the mitigating activities. There are many ways to deal with threats. Senior management is primarily interested in how the security program can mitigate threats while supporting the ultimate business goals. While benchmarking against other enterprises can provide an approach, it does not necessarily mean that the approach is the best one for the enterprise.

S1-127   Alignment of a security program to business objectives is **BEST** achieved through:

     A.   senior management directing the security program.
     B.   periodic risk analysis and treatment.
     C.   a security steering committee with representatives from all business functions.
     D.   regular security audits and ongoing monitoring.

C   To ensure alignment, the security program should establish a steering committee that includes all business areas. This allows all key stakeholders to provide input and make collective decisions regarding security program activities. Senior management establishes the direction for the security program, but the alignment will only happen when the security program is executed to support the business objectives. Risk analysis and treatment are part of the security program's risk management process, but this process may not align with the business objectives. Regular security audits and ongoing monitoring ensure the execution of the security program, but have a limited effect on alignment to the business objectives.

S1-128   Which of the following will have the **GREATEST** impact on a financial enterprise with offices in various countries and involved in transborder flow of information?

     A.   Current and future technologies
     B.   Evolving data protection regulations
     C.   Economizing the costs of network bandwidth
     D.   Centralization of information security

B   Information security laws vary from country to country and an enterprise must be aware of and follow the applicable laws from each country. There are regulations from countries mandating the data security requirements, and these generally should be followed wherever the data are flowing between the various offices. The other choices would be considered, but will have less impact compared to regulatory requirements.

S1-129   Strategic alignment is **PRIMARILY** achieved when services provided by the information security
          department:

          A.   closely reflect the requirements of key business stakeholders.
          B.   closely reflect the desires of the IT executive team.
          C.   reflect the requirements of industry best practices.
          D.   are reliable and cost-effective using the latest technologies.

A   The information security strategic plan should be aligned to the business strategy. Business strategy is
          the articulation of the desires of the business executive team and the board of directors, who are key
          stakeholders. IT strategic alignment is achieved when it closely reflects the requirements and desires of
          these business users. The IT executive team does not necessarily reflect the opinion and requirements of the
          broader business. Choice C is wrong because industry best practices may not be the right solution for the
          business. Even if the solution is reliable and cost-effective, if it does not meet the business needs then it is
          not directed toward business advantage.

S1-130   Who is in the **BEST** position to implement and monitor a balanced scorecard (BSC) for the information
          systems (IS) security program?

          A.   Executive management
          B.   The chief information security officer (CISO)
          C.   The director of auditing
          D.   The chief information officer (CIO)

B   An IT BSC demonstrates IT value, facilitates IT governance, and acts as a decision support tool for
          IT management. The CISO develops, implements and monitors the performance metrics as part of the
          information security governance framework. It is the role of executive management to provide support
          to IS management to implement measures to achieve the security objectives. The director of auditing
          oversees the execution of various audit plans and provides assurance that controls are implemented and
          operating effectively to support the objectives. The CIO is responsible for the technology governance of the
          enterprise.

S1-131   Which of the following is the **MOST** important factor on which to rely to successfully assign
          cross-organizational responsibility to integrate an information security program?

          A.   The ease of information security technologies
          B.   Open channels of communication
          C.   The roles of different job functions
          D.   Qualified information security professionals in each department

C   Job functions across the organization must be taken into consideration before assigning responsibility
          within the information security program. The transparency of information security technologies and
          processes is important at the end-user level to ensure that information security does not reduce the
          efficiency of existing work practices, encouraging work-arounds or other actions that render controls
          ineffective. Open channels of communication are important, but do not necessarily lead to assigning
          responsibility for information security control to another person. Having qualified information security
          professionals in each department will not necessarily translate into a willingness to accept information
          security responsibility.

S1-132   The security responsibility of data custodians in an organization will include:

    A. assuming overall protection of information assets.
    B. determining data classification levels.
    C. implementing security controls in products they install.
    D. ensuring security measures are consistent with policy.

D   Security responsibilities of data custodians within an organization include ensuring that appropriate security measures are maintained and are consistent with organizational policy. Executive management holds overall responsibility for protection of the information assets. Data owners determine data classification levels for information assets so that appropriate levels of controls can be provided to meet the requirements relating to confidentiality, integrity and availability. Implementation of information security in products is the responsibility of the IT developers.

S1-133   Who can **BEST** approve plans to implement an information security governance framework?

    A. Internal auditor
    B. Information security management
    C. Steering committee
    D. Infrastructure management

C   Senior management that is part of the security steering committee is in the best position to approve plans to implement an information security governance framework. An internal auditor is secondary to the authority and influence of senior management. Information security management should not have the authority to approve the security governance framework. Infrastructure management will not be in the best position since it focuses more on the technologies than on the business.

S1-134   What is the **MOST** important item to be included in an information security policy?

    A. The definition of roles and responsibilities
    B. The scope of the security program
    C. The key objectives of the security program
    D. Reference to procedures and standards of the security program

C   Stating the objectives of the security program is the most important element to ensure alignment with business goals. The other choices are part of the security policy, but they are not as important.

S1-135   In an organization, information systems security is the responsibility of:

    A. all personnel.
    B. information systems personnel.
    C. information systems security personnel.
    D. functional personnel.

A   All personnel of the organization have the responsibility of ensuring information systems security—this can include indirect personnel such as physical security personnel. Information systems security cannot be the responsibility of information systems personnel alone since they cannot ensure security. Information systems security cannot be the responsibility of information systems security personnel alone since they cannot ensure security. Information systems security cannot be the responsibility of functional personnel alone since they cannot ensure security.

S1-136   An organization that has decided to implement a formal information security program should **FIRST**:

      A.   invite an external consultant to create the security strategy.
      B.   allocate budget based on best practices.
      C.   benchmark similar organizations.
      D.   define high-level business security requirements.

D   All four choices are valid steps in the process of implementing a formal information security program; however, defining high-level business security requirements should precede the others because the implementation should be based on those security requirements.

S1-137   Which of the following is a key area of the ISO 27001 framework?

      A.   Operational risk assessment
      B.   Financial crime metrics
      C.   Capacity management
      D.   Business continuity management

D   Operational risk assessment, financial crime metrics and capacity management can complement the information security framework, but only business continuity management is a key component.

S1-138   The **MAIN** goal of an information security strategic plan is to:

      A.   develop a risk assessment plan.
      B.   develop a data protection plan.
      C.   protect information assets and resources.
      D.   establish security governance.

C   The main goal of an information security strategic plan is to protect information assets and resources. Developing a risk assessment plan and a data protection plan, and establishing security governance refer to tools utilized in the security strategic plan that achieve the protection of information assets and resources.

S1-139   Information security policies should:

      A.   address corporate network vulnerabilities.
      B.   address the process for communicating a violation.
      C.   be straightforward and easy to understand.
      D.   be customized to specific groups and roles.

C   As high-level statements, information security policies should be straightforward and easy to understand. They are high-level and, therefore, do not address network vulnerabilities directly or the process for communicating a violation. As policies, they should provide a uniform message to all groups and user roles.

S1-140   Requiring all employees and contractors to meet personnel security/suitability requirements commensurate with their position's sensitivity level and subject to personnel screening is an example of a security:

    A. policy.
    B. strategy.
    C. guideline.
    D. baseline.

A   A security policy is a general statement to define management objectives with respect to security. The security strategy addresses higher level issues. Guidelines are optional actions and operational tasks. A security baseline is a set of minimum requirements that is acceptable to an organization.

S1-141   The **MOST** important aspect in establishing good information security policies is to ensure that they:

    A. have the consensus of all concerned groups.
    B. are easy to access by all employees.
    C. capture the intent of management and align with business goals.
    D. have been approved by the internal audit department.

C   Policies should reflect the intent and direction of the business and IT or they will never be adopted. Having the consensus of all concerned groups will never happen. Availability of policies is important, but not an indicator of good information security content. The internal audit department ensures that policies are being followed, but it does not write the policies.

DOMAIN 2—INFORMATION RISK MANAGEMENT AND COMPLIANCE (33%)

S2-1   An effective risk management program should reduce risk to:

A. zero.
B. an acceptable level.
C. an acceptable percent of revenue.
D. an acceptable probability of occurrence.

B   Risk should be reduced to an acceptable level based on the risk preference of the organization. Reducing risk to zero is impractical and could be cost-prohibitive. Tying risk to a percentage of revenue is inadvisable since there is no direct correlation between the two. Reducing the probability of risk occurrence may not always be possible, as in the case of natural disasters. The focus should be on reducing the impact to an acceptable level to the organization, not reducing the probability of the risk.

S2-2   The **MOST** important reason for conducting periodic risk assessments is because:

A. risk assessments are not always precise.
B. security risks are subject to frequent change.
C. reviewers can optimize and reduce the cost of controls.
D. it demonstrates to senior management that the security function can add value.

B   Risks are constantly changing. A previously conducted risk assessment may not include measured risks that have been introduced since the last assessment. Although an assessment can never be perfect and invariably contains some errors, this is not the most important reason for periodic reassessment. The fact that controls can be made more efficient to reduce costs is not sufficient. Finally, risk assessments should not be performed merely to justify the existence of the security function.

S2-3   Which of the following **BEST** indicates a successful risk management practice?

A. Overall risk is quantified
B. Inherent risk is eliminated
C. Residual risk is minimized
D. Control risk is tied to business units

C   A successful risk management practice minimizes the residual risk to the organization. Choice A is incorrect because the fact that overall risk has been quantified does not necessarily indicate the existence of a successful risk management practice. Choice B is incorrect since it is virtually impossible to eliminate inherent risk. Choice D is incorrect because, although the tying of control risks to business may improve accountability, this is not as desirable as minimizing residual risk.

S2-4   A successful information security management program should use which of the following to determine the amount of resources devoted to mitigating exposures?

A. Risk analysis results
B. Audit report findings
C. Penetration test results
D. Amount of IT budget available

A   Risk analysis results are the most useful and complete source of information for determining the amount of resources to devote to mitigating exposures. Audit report findings may not address all risks and do not address annual loss frequency. Penetration test results provide only a limited view of exposures, while the IT budget is not tied to the exposures faced by the organization.

S2-5   Which of the following will **BEST** protect an organization from internal security attacks?

    A. Static IP addressing
    B. Internal address translation
    C. Prospective employee background checks
    D. Employee awareness certification program

C   Because past performance is a strong predictor of future performance, background checks of prospective employees best prevents attacks from originating within an organization. Static IP addressing does little to prevent an internal attack. Internal address translation using nonroutable addresses is useful against external attacks but not against internal attacks. Employees who certify that they have read security policies are desirable, but this does not guarantee that the employees behave honestly.

S2-6   For risk management purposes, the value of a physical asset should be based on:

    A. original cost.
    B. net cash flow.
    C. net present value.
    D. replacement cost.

D   The value of a physical asset should be based on its replacement cost since this is the amount that would be needed to replace the asset if it were to become damaged or destroyed. Original cost may be significantly different than the current cost of replacing the asset. Net cash flow and net present value do not accurately reflect the true value of the asset.

S2-7   In a business impact analysis, the value of an information system should be based on the overall cost:

    A. of recovery.
    B. to recreate.
    C. if unavailable.
    D. of emergency operations.

C   The value of an information system should be based on the cost incurred if the system were to become unavailable. The cost to design or recreate the system is not as relevant since a business impact analysis measures the impact that would occur if an information system were to become unavailable. Similarly, the cost of emergency operations is not as relevant.

S2-8   The value of information assets is **BEST** determined by:

    A. individual business managers.
    B. business systems analysts.
    C. information security management.
    D. industry averages benchmarking.

A   Individual business managers are in the best position to determine the value of information assets since they are most knowledgeable of the assets' impact on the business. Business systems developers and information security managers are not as knowledgeable regarding the impact on the business. Peer companies' industry averages do not necessarily provide detailed enough information nor are they as relevant to the unique aspects of the business.

S2-9   During which phase of development is it **MOST** appropriate to begin assessing the risk of a new application system?

    A. Feasibility
    B. Design
    C. Development
    D. Testing

A   Risk should be addressed as early in the development of a new application system as possible. In some cases, identified risks could be mitigated through design changes. If needed changes are not identified until design has already commenced, such changes become more expensive. For this reason, beginning risk assessment during the design, development or testing phases is not the best solution.

S2-10   The **MOST** effective way to incorporate risk management practices into existing production systems is through:

    A. policy development.
    B. change management.
    C. awareness training.
    D. regular monitoring.

B   Change is a process in which new risks can be introduced into business processes and systems. For this reason, risk management should be an integral component of the change management process. Policy development, awareness training and regular monitoring, although all worthwhile activities, are not as effective as change management.

S2-11   Which of the following would be **MOST** useful in developing a series of recovery time objectives (RTOs)?

    A. Gap analysis
    B. Regression analysis
    C. Risk analysis
    D. Business impact analysis

D   Recovery time objectives (RTOs) are a primary deliverable of a business impact analysis. RTOs relate to the financial impact of a system not being available. A gap analysis is useful in addressing the differences between the current state and an ideal future state. Regression analysis is used to test changes to program modules. Risk analysis is a component of the business impact analysis.

S2-12   The decision on whether new risks should fall under periodic or event-driven reporting should be based on which of the following?

    A. Mitigating controls
    B. Visibility of impact
    C. Likelihood of occurrence
    D. Incident frequency

B   Visibility of impact is the best measure since it manages risks to an organization in the timeliest manner. Likelihood of occurrence and incident frequency are not as relevant. Mitigating controls is not a determining factor on incident reporting.

S2-13   Risk acceptance is a component of which of the following?

          A. Assessment
          B. Treatment
          C. Evaluation
          D. Monitoring

B   Risk acceptance is one of the alternatives to be considered in the risk treatment process. Assessment and evaluation are components of the risk analysis process. Risk acceptance is not a component of monitoring.

S2-14   Risk management programs are designed to reduce risk to:

          A. a level that is too small to be measurable.
          B. the point at which the benefit exceeds the expense.
          C. a level that the organization is willing to accept.
          D. a rate of return that equals the current cost of capital.

C   Risk should be reduced to a level that an organization is willing to accept. Reducing risk to a level too small to measure is impractical and is often cost-prohibitive. To tie risk to a specific rate of return ignores the qualitative aspects of risk that must also be considered. Depending on the risk preference of an organization, it may or may not choose to pursue risk mitigation to the point at which the benefit equals or exceeds the expense. Therefore, choice C is a more precise answer.

S2-15   A risk assessment should **TYPICALLY** be conducted:

          A. once a year for each business process and subprocess.
          B. every three to six months for critical business processes.
          C. by external parties to maintain objectivity.
          D. annually or whenever there is a significant change.

D   Risks are constantly changing. Choice D offers the best alternative because it takes into consideration a reasonable time frame and allows flexibility to address significant change. Conducting a risk assessment once a year is insufficient if important changes take place. Conducting a risk assessment every three-to-six months for critical processes may not be necessary, or it may not address important changes in a timely manner. It is not necessary for assessments to be performed by external parties.

S2-16   Which of the following risks would **BEST** be assessed using qualitative risk assessment techniques?

          A. Theft of purchased software
          B. Power outage lasting 24 hours
          C. Permanent decline in customer confidence
          D. Temporary loss of e-mail due to a virus attack

C   A permanent decline in customer confidence does not lend itself well to measurement by quantitative techniques. Qualitative techniques are more effective in evaluating things such as customer loyalty and goodwill. Theft of software, power outages and temporary loss of e-mail can be quantified into monetary amounts easier than can be assessed with quantitative techniques.

S2-17   A business impact analysis (BIA) is the **BEST** tool for calculating:

    A. total cost of ownership.
    B. priority of restoration.
    C. annualized loss expectancy (ALE).
    D. residual risk.

B   A business impact analysis (BIA) is the best tool for calculating the priority of restoration for applications. It is not used to determine total cost of ownership, annualized loss expectancy (ALE) or residual risk to the organization.

S2-18   Quantitative risk analysis is **MOST** appropriate when assessment data:

    A. include customer perceptions.
    B. contain percentage estimates.
    C. do not contain specific details.
    D. contain subjective information.

B   Percentage estimates are characteristic of quantitative risk analysis. Customer perceptions, lack of specific details or subjective information lend themselves more to qualitative risk analysis.

S2-19   Which of the following is the **MOST** appropriate use of gap analysis?

    A. Evaluating a business impact analysis (BIA)
    B. Developing a balanced business scorecard
    C. Demonstrating the relationship between controls
    D. Measuring current state vs. desired future state

D   A gap analysis is most useful in addressing the differences between the current state and an ideal future state. It is not as appropriate for evaluating a business impact analysis (BIA), developing a balanced business scorecard or demonstrating the relationship between variables.

S2-20   A risk analysis should:

    A. include a benchmark of similar companies in its scope.
    B. assume an equal degree of protection for all assets.
    C. address the potential size and likelihood of loss.
    D. give more weight to the likelihood vs. the size of the loss.

C   A risk analysis should take into account the potential size and likelihood of a loss. It could include comparisons with a group of companies of similar size. It should not assume an equal degree of protection for all assets since assets may have different risk factors. The likelihood of the loss should not receive greater emphasis than the size of the loss; a risk analysis should always address both equally.

S2-21   Based on the information provided, which of the following situations presents the **GREATEST** information security risk for an organization with multiple, but small, domestic processing locations?

    A. Systems operation guidelines are not enforced
    B. Change management procedures are poor
    C. Systems development is outsourced
    D. Systems capacity management is not performed

B   The lack of change management is a severe omission and will greatly increase information security risk. Since guidelines are generally nonauthoritative, their lack of enforcement is not a primary concern. Systems that are developed by third-party vendors are becoming commonplace and do not represent an increase in security risk as much as poor change management. Poor capacity management may not necessarily represent a security risk.

S2-22   Which of the following **BEST** describes the scope of risk analysis?

    A   Key financial systems
    B. Organizational activities
    C. Key systems and infrastructure
    D. Systems subject to regulatory compliance

B   Risk analysis should include all organizational activities. It should not be limited to subsets of systems or just systems and infrastructure.

S2-23   The decision as to whether a risk has been reduced to an acceptable level should be determined by:

    A. organizational requirements.
    B. information systems requirements.
    C. information security requirements.
    D. international standards.

A   Organizational requirements should determine when a risk has been reduced to an acceptable level. Information systems and information security should not make the ultimate determination. Since each organization is unique, international standards of best practice do not represent the best solution.

S2-24   Which of the following is the **PRIMARY** reason for implementing a risk management program?

    A. Allows the organization to eliminate risk
    B. Is a necessary part of management's due diligence
    C. Satisfies audit and regulatory requirements
    D. Assists in incrementing the return on investment (ROI)

B   The key reason for performing risk management is that it is part of management's due diligence. The elimination of all risk is not possible. Satisfying audit and regulatory requirements is of secondary importance. A risk management program may or may not increase the return on investment (ROI).

S2-25 Which of the following groups would be in the **BEST** position to perform a risk analysis for a business?

    A. External auditors
    B. A peer group within a similar business
    C. Process owners
    D. A specialized management consultant

C   Process owners have the most in-depth knowledge of risks and compensating controls within their environment. External parties do not have that level of detailed knowledge on the inner workings of the business. Management consultants are expected to have the necessary skills in risk analysis techniques but are still less effective than a group with intimate knowledge of the business.

S2-26 A successful risk management program should lead to:

    A. optimization of risk reduction efforts against cost.
    B. containment of losses to an annual budgeted amount.
    C. identification and removal of all man-made threats.
    D. elimination or transference of all organizational risks.

A   Successful risk management should lead to a breakeven point of risk reduction and cost. The other options listed are not achievable. Threats cannot be totally removed or transferred, while losses cannot be budgeted in advance with absolute certainty.

S2-27 Which of the following risks would **BEST** be assessed using quantitative risk assessment techniques?

    A. Customer data stolen
    B. An electrical power outage
    C. A web site defaced by hackers
    D. Loss of the software development team

B   The effect of the theft of customer data or web site defacement by hackers could lead to a permanent decline in customer confidence, which does not lend itself to measurement by quantitative techniques. Loss of a majority of the software development team could have similar unpredictable repercussions. However, the loss of electrical power for a short duration is more easily measurable and can be quantified into monetary amounts that can be assessed with quantitative techniques.

S2-28 The impact of losing frame relay network connectivity for 18-24 hours should be calculated using the:

    A. hourly billing rate charged by the carrier.
    B. value of the data transmitted over the network.
    C. aggregate compensation of all affected business users.
    D. financial losses incurred by affected business units.

D   The bottom line on calculating the impact of a loss is what its cost will be to the organization. The other choices are all factors that contribute to the overall monetary impact.

S2-29   Which of the following is the **MOST** usable deliverable of an information security risk analysis?

    A. Business impact analysis (BIA) report
    B. List of action items to mitigate risk
    C. Assignment of risks to process owners
    D. Quantification of organizational risk

B   Although all of these are important, the list of action items is used to reduce or transfer the current level of risk. The other options materially contribute to the way the actions are implemented.

S2-30   Ongoing tracking of remediation efforts to mitigate identified risks can **BEST** be accomplished through the use of which of the following?

    A. Tree diagrams
    B. Venn diagrams
    C. Heat charts
    D. Bar charts

C   Heat charts, sometimes referred to as stoplight charts, quickly and clearly show the current status of remediation efforts. Venn diagrams show the connection between sets; tree diagrams are useful for decision analysis; and bar charts show relative size.

S2-31   Which two components **PRIMARILY** must be assessed in an effective risk analysis?

    A. Visibility and duration
    B. Likelihood and impact
    C. Probability and frequency
    D. Financial impact and duration

B   The probability or likelihood of the event and the financial impact or magnitude of the event must be assessed first. Duration refers to the length of the event; it is important in order to assess impact but is secondary. Once the likelihood is determined, the frequency is also important to determine overall impact.

S2-32   Information security managers should use risk assessment techniques to:

    A. justify selection of risk mitigation strategies.
    B. maximize the return on investment (ROI).
    C. provide documentation for auditors and regulators.
    D. quantify risks that would otherwise be subjective.

A   Information security managers should use risk assessment techniques to justify and implement a risk mitigation strategy as efficiently as possible. None of the other choices accomplishes that task, although they are important components.

S2-33   In assessing risk, it is **MOST** essential to:

A. provide equal coverage for all asset types.
B. use benchmarking data from similar organizations.
C. consider both monetary value and likelihood of loss.
D. focus primarily on threats and recent business losses.

C   A risk analysis should take into account the potential financial impact and likelihood of a loss. It should not weigh all potential losses evenly, nor should it focus primarily on recent losses or losses experienced by similar firms. Although this is important supplementary information, it does not reflect the organization's real situation. Geography and other factors come into play as well.

S2-34   Data owners are **PRIMARILY** responsible for establishing risk mitigation methods to address which of the following areas?

A. Platform security
B. Entitlement changes
C. Intrusion detection
D. Antivirus controls

B   Data owners are responsible for assigning user entitlements and approving access to the systems for which they are responsible. Platform security, intrusion detection and antivirus controls are all within the responsibility of the information security manager.

S2-35   The **PRIMARY** goal of a corporate risk management program is to ensure that an organization's:

A. IT assets in key business functions are protected.
B. business risks are addressed by preventive controls.
C. stated objectives are achievable.
D. IT facilities and systems are always available.

C   Risk management's primary goal is to ensure an organization maintains the ability to achieve its objectives. Protecting IT assets is one possible goal as well as ensuring infrastructure and systems availability. However, these should be put in the perspective of achieving an organization's objectives. Preventive controls are not always possible or necessary; risk management will address issues with an appropriate mix of preventive and corrective controls.

S2-36   It is important to classify and determine relative sensitivity of assets to ensure that:

A. cost of protection is in proportion to sensitivity.
B. highly sensitive assets are protected.
C. cost of controls is minimized.
D. countermeasures are proportional to risk.

D   Classification of assets needs to be undertaken to determine sensitivity of assets in terms of risk to the business operation so that proportional countermeasures can be effectively implemented. While higher costs are allowable to protect sensitive assets, and it is always reasonable to minimize the costs of controls, it is most important that the controls and countermeasures are commensurate to the risk since this will justify the costs. Choice B is important but it is an incomplete answer because it does not factor in risk. Therefore, choice D is the most important.

S2-37   The service level agreement (SLA) for an outsourced IT function does not reflect an adequate level of protection. In this situation an information security manager should:

A. ensure the provider is made liable for losses.
B. recommend not renewing the contract upon expiration.
C. recommend the immediate termination of the contract.
D. determine the current level of security.

D   It is important to ensure that adequate levels of protection are written into service level agreements (SLAs) and other outsourcing contracts. Information must be obtained from providers to determine how that outsource provider is securing information assets prior to making any recommendation or taking any action in order to support management decision making. Choice A is not acceptable in most situations and therefore not a good answer.

S2-38   An information security manager has been assigned to implement more restrictive preventive controls. By doing so, the net effect will be to **PRIMARILY** reduce the:

A. threat.
B. loss.
C. vulnerability.
D. probability.

C   Implementing more restrictive preventive controls mitigates vulnerabilities but not the threats. Losses and probability of occurrence may not be primarily or directly affected.

S2-39   When performing a quantitative risk analysis, which of the following is **MOST** important to estimate the potential loss?

A. Evaluate productivity losses
B. Assess the impact of confidential data disclosure
C. Calculate the value of the information or asset
D. Measure the probability of occurrence of each threat

C   Calculating the value of the information or asset is the first step in a risk analysis process to determine the impact to the organization, which is the ultimate goal. Determining how much productivity could be lost and how much it would cost is a step in the estimation of potential risk process. Knowing the impact if confidential information is disclosed is also a step in the estimation of potential risk. Measuring the probability of occurrence for each threat identified is a step in performing a threat analysis and therefore a partial answer.

S2-40   Before conducting a formal risk assessment of an organization's information resources, an information security manager should **FIRST**:

A. map the major threats to business objectives.
B. review available sources of risk information.
C. identify the value of the critical assets.
D. determine the financial impact if threats materialize.

A   Risk mapping or a macro assessment of the major threats to the organization is a simple first step before performing a risk assessment. Compiling all available sources of risk information is part of the risk assessment. Choices C and D are also components of the risk assessment process, which are performed subsequent to the threats-business mapping.

S2-41   The valuation of IT assets should be performed by:

A. an IT security manager.
B. an independent security consultant.
C. the chief financial officer (CFO).
D. the information owner.

D   Information asset owners are in the best position to evaluate the value added by the IT asset under review within a business process, thanks to their deep knowledge of the business processes and of the functional IT requirements. An IT security manager is an expert of the IT risk assessment methodology and IT asset valuation mechanisms. However, the manager could not have a deep understanding of all the business processes of the firm. An IT security subject matter expert will take part of the process to identify threats and vulnerabilities and will collaborate with the business information asset owner to define the risk profile of the asset. A chief financial officer (CFO) will have an overall costs picture but not detailed enough to evaluate the value of each IT asset.

S2-42   The **PRIMARY** objective of a risk management program is to:

A. minimize inherent risk.
B. eliminate business risk.
C. implement effective controls.
D. reduce residual risk to acceptable levels.

D   The goal of a risk management program is to ensure that residual risk is reduced and remains at acceptable levels. Management of risk does not always require the removal of inherent risk nor is this always possible. Elimination of business risk is not possible. Effective controls are naturally a clear objective of a risk management program, but with the choices given, Choice C is an incomplete answer.

S2-43   After completing a full IT risk assessment, who will **BEST** decide which mitigating controls should be implemented?

A. Senior management
B. Business manager
C. IT audit manager
D. Information security officer (ISO)

B   The business manager will be in the best position, based on the risk assessment and mitigation proposals, to decide which controls should/could be implemented, in line with the business strategy and with budget. Senior management will have to ensure that the business manager has a clear understanding of the risk assessed but in no case will be in a position to decide on specific controls. The IT audit manager will take part in the process to identify threats and vulnerabilities, and to make recommendations for mitigations. The information security officer (ISO) could make some decisions regarding implementation of controls. However, the business manager will have a broader business view and full control over the budget and, therefore, will be in a better position to make strategic decisions.

S2-44   When performing an information risk analysis, an information security manager should **FIRST**:

A. establish the ownership of assets.
B. evaluate the risks to the assets.
C. take an asset inventory.
D. categorize the assets.

C   Assets must be inventoried before any of the other choices can be performed.

CISM Review Questions, Answers & Explanations Manual 2012   **53**
ISACA. All Rights Reserved.

S2-45   The **PRIMARY** benefit of performing an information asset classification is to:

A. link security requirements to business objectives.
B. identify controls commensurate to risk.
C. define access rights.
D. establish ownership.

B   All choices are benefits of information classification. However, identifying controls that are proportional to the risk in all cases is the primary benefit of the process.

S2-46   Which of the following is **MOST** essential for a risk management program to be effective?

A. Flexible security budget
B. Sound risk baseline
C. Detection of new risk
D. Accurate risk reporting

C   All of these procedures are essential for implementing risk management. However, without identifying new risk, other procedures will only be useful for a limited period.

S2-47   Which of the following attacks is **BEST** mitigated by utilizing strong passwords?

A. Man-in-the-middle attack
B. Brute force attack
C. Remote buffer overflow
D. Root kit

B   A brute force attack is normally successful against weak passwords, whereas strong passwords would not prevent any of the other attacks. Man-in-the-middle attacks intercept network traffic, which could contain passwords, but is not naturally password-protected. Remote buffer overflows rarely require a password to exploit a remote host. Root kits hook into the operating system's kernel and, therefore, operate underneath any authentication mechanism.

S2-48   Phishing is **BEST** mitigated by which of the following?

A. Security monitoring software
B. Encryption
C. Two-factor authentication
D. User awareness

D   Phishing can best be detected by the user. It can be mitigated by appropriate user awareness. Security monitoring software would provide some protection, but would not be as effective as user awareness. Encryption and two-factor authentication would not mitigate this threat.

S2-49   Risk assessments should be repeated at regular intervals because:

        A. business threats are constantly changing.
        B. omissions in earlier assessments can be addressed.
        C. repetitive assessments allow various methodologies.
        D. they help raise awareness on security in the business.

A   As business objectives and methods change, the nature and relevance of threats change as well. Choice B does not, by itself, justify regular reassessment. Choice C is not necessarily true in all cases. Choice D is incorrect because there are better ways of raising security awareness than by performing a risk assessment.

S2-50   Which of the following steps in conducting a risk assessment should be performed **FIRST**?

        A. Identify business assets
        B. Identify business risks
        C. Assess vulnerabilities
        D. Evaluate key controls

A   Risk assessment first requires one to identify the business assets that need to be protected before identifying the threats. The next step is to establish whether those threats represent business risk by identifying the likelihood and effect of occurrence, followed by assessing the vulnerabilities that may affect the security of the asset. This process establishes the control objectives against which key controls can be evaluated.

S2-51   A risk management program would be expected to:

        A. remove all inherent risk.
        B. maintain residual risk at an acceptable level.
        C. implement preventive controls for every threat.
        D. reduce control risk to zero.

B   The object of risk management is to ensure that all residual risk is maintained at a level acceptable to the business; it is not intended to remove every identified risk or implement controls for every threat since this may not be cost-effective. Control risk, i.e., that a control may not be effective, is a component of the program but is unlikely to be reduced to zero.

S2-52   In which phase of the development process should risk assessment be **FIRST** introduced?

        A. Programming
        B. Specification
        C. User testing
        D. Feasibility

D   Risk should be addressed as early as possible in the development cycle. The feasibility study should include risk assessment so that the cost of controls can be estimated before the project proceeds. Risk should also be considered in the specification phase where the controls are designed, but this would still be based on the assessment carried out in the feasibility study. Assessment would not be relevant in choice A or C.

S2-53   Which of the following would help management determine the resources needed to mitigate a risk to the organization?

A.   Risk analysis process
B.   Business impact analysis (BIA)
C.   Risk management balanced scorecard
D.   Risk-based audit program

B   The business impact analysis (BIA) determines the possible outcome of a risk and is essential to determine the appropriate cost of control. The risk analysis process provides comprehensive data, but does not determine definite resources to mitigate the risk as does the BIA. The risk management balanced scorecard is a measuring tool for goal attainment. A risk-based audit program is used to focus the audit process on the areas of greatest importance to the organization.

S2-54   A global financial institution has decided not to take any further action on a denial of service (DoS) risk found by the risk assessment team. The **MOST** likely reason they made this decision is that:

A.   there are sufficient safeguards in place to prevent this risk from happening.
B.   the needed countermeasure is too complicated to deploy.
C.   the cost of countermeasure outweighs the value of the asset and potential loss.
D.   The likelihood of the risk occurring is unknown.

C   An organization may decide to live with specific risks because it would cost more to protect themselves than the value of the potential loss. The safeguards need to match the risk level. While countermeasures could be too complicated to deploy, this is not the most compelling reason. It is unlikely that a global financial institution would not be exposed to such attacks and the frequency could not be predicted.

S2-55   Which of the following types of information would the information security manager expect to have the **LOWEST** level of security protection in a publicly traded, multinational enterprise?

A.   Strategic business plan
B.   Upcoming financial results
C.   Customer personal information
D.   Previous financial results

D   Previous financial results are public; all of the other choices are private information and should only be accessed by authorized entities.

S2-56   The **PRIMARY** purpose of using risk analysis within a security program is to:

A.   justify the security expenditure.
B.   help businesses prioritize the assets to be protected.
C.   inform executive management of residual risk value.
D.   assess exposures and plan remediation.

D   Risk analysis explores the degree to which an asset needs protecting so this can be managed effectively. Risk analysis indirectly supports the security expenditure, but justifying the security expenditure is not its primary purpose. Helping businesses prioritize the assets to be protected is an indirect benefit of risk analysis, but not its primary purpose. Informing executive management of residual risk value is not directly relevant.

S2-57   Which of the following is the **PRIMARY** prerequisite to implementing data classification within
         an organization?

         A. Defining job roles
         B. Performing a risk assessment
         C. Identifying data owners
         D. Establishing data retention policies

C   Identifying the data owners is the first step, and is essential to implementing data classification. Defining
         job roles is not relevant. Performing a risk assessment is important, but will require the participation of
         data owners (who must first be identified). Establishing data retention policies may occur after data have
         been classified.

S2-58   An online banking institution is concerned that the breach of customer personal information will have
         a significant financial impact due to the need to notify and compensate customers whose personal
         information may have been compromised. The institution determines that residual risk will always be too
         high and decides to:

         A. mitigate the impact by purchasing insurance.
         B. implement a circuit-level firewall to protect the network.
         C. increase the resiliency of security measures in place.
         D. implement a real-time intrusion detection system.

A   Since residual risk will always be too high, the only practical solution is to mitigate the financial impact by
         purchasing insurance. Purchasing insurance is also known as risk transference.

S2-59   What mechanisms are used to identify deficiencies that would provide attackers with an opportunity to
         compromise a computer system?

         A. Business impact analyses
         B. Security gap analyses
         C. System performance metrics
         D. Incident response processes

B   A security gap analysis is a process which measures all security controls in place against typically good
         business practice, and identifies related weaknesses. A business impact analysis is less suited to identify
         security deficiencies. System performance metrics may indicate security weaknesses, but that is not their
         primary purpose. Incident response processes exist for cases where security weaknesses are exploited.

S2-60   A common concern with poorly written web applications is that they can allow an attacker to:

         A. gain control through a buffer overflow.
         B. conduct a distributed denial of service (DoS) attack.
         C. abuse a race condition.
         D. inject structured query language (SQL) statements.

D   Structured query language (SQL) injection is one of the most common and dangerous web application
         vulnerabilities. Buffer overflows and race conditions are very difficult to find and exploit on web applications.
         Distributed denial of service (DoS) attacks have nothing to do with the quality of a web application.

S2-61   Which of the following would be of **GREATEST** importance to the security manager in determining
         whether to further mitigate residual risk?

         A. Historical cost of the asset
         B. Acceptable level of potential business impacts
         C. Cost versus benefit of additional mitigating controls
         D. Annualized loss expectancy (ALE)

C   The security manager would be most concerned with whether residual risk would be reduced by a greater
         amount than the cost of adding additional controls. The other choices, although relevant, would not be
         as important.

S2-62   A project manager is developing a developer portal and requests that the security manager assign a public
         IP address so that it can be accessed by in-house staff and by external consultants outside the organization's
         local are network (LAN). What should the security manager do **FIRST**?

         A. Understand the business requirements of the developer portal
         B. Perform a vulnerability assessment of the developer portal
         C. Install an intrusion detection system (IDS)
         D. Obtain a signed nondisclosure agreement (NDA) from the external consultants before allowing external
             access to the server

A   The information security manager cannot make an informed decision about the request without first
         understanding the business requirements of the developer portal. Performing a vulnerability assessment of
         developer portal and installing an intrusion detection system (IDS) are best practices but are subsequent to
         understanding the requirements. Obtaining a signed nondisclosure agreement will not take care of the risks
         inherent in the organization's application.

S2-63   A mission-critical system has been identified as having an administrative system account with attributes
         that prevent locking and change of privileges and name. Which would be the **BEST** approach to prevent
         successful brute forcing of the account?

         A. Prevent the system from being accessed remotely
         B   Create a strong random password
         C. Ask for a vendor patch
         D. Track usage of the account by audit trails

B   Creating a strong random password reduces the risk of a successful brute force attack by exponentially
         increasing the time required. Preventing the system from being accessed remotely is not always an option
         in mission-critical systems and still leaves local access risks. Vendor patches are not always available.
         Tracking usage is a detective control and will not prevent an attack.

S2-64   Attackers who exploit cross-site scripting vulnerabilities take advantage of:

     A.   a lack of proper input validation controls.
     B.   weak authentication controls in the web application layer.
     C.   flawed cryptographic secure sockets layer (SSL) implementations and short key lengths.
     D.   implicit web application trust relationships.

A   Cross-site scripting attacks inject malformed input. Attackers who exploit weak application authentication controls can gain unauthorized access to applications and this has little to do with cross-site scripting vulnerabilities. Attackers who exploit flawed cryptographic secure sockets layer (SSL) implementations and short key lengths can sniff network traffic and crack keys to gain unauthorized access to information. This has little to do with cross-site scripting vulnerabilities. Web application trust relationships do not relate directly to the attack.

S2-65   Which of the following would **BEST** address the risk of data leakage?

     A.   File backup procedures
     B.   Database integrity checks
     C.   Acceptable use policies
     D.   Incident response procedures

C   Acceptable use policies are the best measure for preventing the unauthorized disclosure of confidential information. The other choices do not address confidentiality of information.

S2-66   A company recently developed a breakthrough technology. Since this technology could give this company a significant competitive edge, which of the following would **FIRST** govern how this information is to be protected?

     A.   Access control policy
     B.   Data classification policy
     C.   Encryption standards
     D.   Acceptable use policy

B   Data classification policies define the level of protection to be provided for each category of data. Without this mandated ranking of degree of protection, it is difficult to determine what access controls or levels of encryption should be in place. An acceptable use policy is oriented more toward the end user and, therefore, would not specifically address what controls should be in place to adequately protect information.

S2-67   What is the **BEST** technique to determine which security controls to implement with a limited budget?

     A.   Risk analysis
     B.   Annualized loss expectancy (ALE) calculations
     C.   Cost-benefit analysis
     D.   Impact analysis

C   Cost-benefit analysis is performed to ensure that the cost of a safeguard does not outweigh it's benefit and that the best safeguard is provided for the cost of implementation. Risk analysis identifies the risks and suggests appropriate mitigation. The annualized loss expectancy (ALE) is a subset of a cost-benefit analysis. Impact analysis would indicate how much could be lost if a specific threat occurred.

S2-68   A company's mail server allows anonymous file transfer protocol (FTP) access which could be exploited. What process should the information security manager deploy to determine the necessity for remedial action?

A.   A penetration test
B.   A security baseline review
C.   A risk assessment
D.   A business impact analysis (BIA)

C   A risk assessment will identify the business impact of such vulnerability being exploited and is, thus, the correct process. A penetration test or a security baseline review may identify the vulnerability but not the remedy. A business impact analysis (BIA) will more likely identify the impact of the loss of the mail server.

S2-69   Which of the following measures would be **MOST** effective against insider threats to confidential information?

A.   Role-based access control
B.   Audit trail monitoring
C.   Privacy policy
D.   Defense-in-depth

A   Role-based access control provides access according to business needs; therefore, it reduces unnecessary access rights and enforces accountability. Audit trail monitoring is a detective control, which is 'after the fact.' Privacy policy is not relevant to this risk. Defense-in-depth primarily focuses on external threats.

S2-70   Because of its importance to the business, an organization wants to quickly implement a technical solution which deviates from the company's policies. An information security manager should:

A.   conduct a risk assessment and allow or disallow based on the outcome.
B.   recommend a risk assessment and implementation only if the residual risks are accepted.
C.   recommend against implementation because it violates the company's policies.
D.   recommend revision of current policy.

B   Whenever the company's policies cannot be followed, a risk assessment should be conducted to clarify the risks. It is then up to management to accept the risks or to mitigate them. Management determines the level of risk they are willing to take. Recommending revision of current policy should not be triggered by a single request.

S2-71   After a risk assessment study, a bank with global operations decided to continue doing business in certain regions of the world where identity theft is rampant. The information security manager should encourage the business to:

A.   increase its customer awareness efforts in those regions.
B.   implement monitoring techniques to detect and react to potential fraud.
C.   outsource credit card processing to a third party.
D.   make the customer liable for losses if they fail to follow the bank's advice.

B   While customer awareness will help mitigate the risks, this is insufficient on its own to control fraud risk. Implementing monitoring techniques which will detect and deal with potential fraud cases is the most effective way to deal with this risk. If the bank outsources its processing, the bank still retains liability. While making the customer liable for losses is a possible approach, nevertheless, the bank needs to be seen to be proactive in managing its risks.

S2-72   The criticality and sensitivity of information assets is determined on the basis of:

A.   threat assessment.
B.   vulnerability assessment.
C.   resource dependency assessment.
D.   impact assessment.

D   The criticality and sensitivity of information assets depends on the impact of the probability of the threats exploiting vulnerabilities in the asset, and takes into consideration the value of the assets and the impairment of the value. Threat assessment lists only the threats that the information asset is exposed to. It does not consider the value of the asset and impact of the threat on the value. Vulnerability assessment lists only the vulnerabilities inherent in the information asset that can attract threats. It does not consider the value of the asset and the impact of perceived threats on the value. Resource dependency assessment provides process needs but not impact.

S2-73   Which program element should be implemented **FIRST** in asset classification and control?

A.   Risk assessment
B.   Classification
C.   Valuation
D.   Risk mitigation

C   Valuation is performed first to identify and understand the assets needing protection. Risk assessment is performed to identify and quantify threats to information assets that are selected by the first step, valuation. Classification and risk mitigation are steps following valuation.

S2-74   When performing a risk assessment, the **MOST** important consideration is that:

A.   management supports risk mitigation efforts.
B.   annual loss expectations (ALEs) have been calculated for critical assets.
C.   assets have been identified and appropriately valued.
D.   attack motives, means and opportunities be understood.

C   Identification and valuation of assets provides the basis for risk management efforts as it relates to the criticality and sensitivity of assets. Management support is always important, but is not relevant when determining the proportionality of risk management efforts. ALE calculations are only valid if assets have first been identified and appropriately valued. Motives, means and opportunities should already be factored in as a part of a risk assessment.

S2-75   The **MAIN** reason why asset classification is important to a successful information security program is because classification determines:

A.   the priority and extent of risk mitigation efforts.
B.   the amount of insurance needed in case of loss.
C.   the appropriate level of protection to the asset.
D.   how protection levels compare to peer organizations.

C   Protection should be proportional to the value of the asset. Classification is based upon the value of the asset to the organization. The amount of insurance needed in case of loss may not be applicable in each case. Peer organizations may have different classification schemes for their assets.

S2-76   The **BEST** strategy for risk management is to:

    A. achieve a balance between risk and organizational goals.
    B. reduce risk to an acceptable level.
    C. ensure that policy development properly considers organizational risks.
    D. ensure that all unmitigated risks are accepted by management.

B   The best strategy for risk management is to reduce risk to an acceptable level, as this will take into account the organization's appetite for risk and the fact that it would not be practical to eliminate all risk. Achieving balance between risk and organizational goals is not always practical. Policy development must consider organizational risks as well as business objectives. It may be prudent to ensure that management understands and accepts risks that it is not willing to mitigate, but that is a practice and is not sufficient to be considered a strategy.

S2-77   Which of the following would be the **MOST** important factor to be considered in the loss of mobile equipment with unencrypted data?

    A. Disclosure of personal information
    B. Sufficient coverage of the insurance policy for accidental losses
    C. Potential impact of the data loss
    D. Replacement cost of the equipment

C   When mobile equipment is lost or stolen, the information contained on the equipment matters most in determining the impact of the loss. The more sensitive the information, the greater the liability. If staff carry mobile equipment for business purposes, an organization must develop a clear policy as to what information should be kept on the equipment and for what purpose. Personal information is not defined in the question as the data that were lost. Insurance may be a relatively smaller issue as compared with information theft or opportunity loss, although insurance is also an important factor for a successful business. Cost of equipment would be a less important issue as compared with other choices.

S2-78   An organization has to comply with recently published industry regulatory requirements—compliance that potentially has high implementation costs. What should the information security manager do **FIRST**?

    A. Implement a security committee.
    B. Perform a gap analysis.
    C. Implement compensating controls.
    D. Demand immediate compliance.

B   Since they are regulatory requirements, a gap analysis would be the first step to determine the level of compliance already in place. Implementing a security committee or compensating controls would not be the first step. Demanding immediate compliance would not assess the situation.

S2-79   Which of the following would be **MOST** relevant to include in a cost-benefit analysis of a two-factor authentication system?

             A.   Annual loss expectancy (ALE) of incidents
             B.   Frequency of incidents
             C.   Total cost of ownership (TCO)
             D.   Approved budget for the project

C   The total cost of ownership (TCO) would be the most relevant piece of information in that it would establish a cost baseline and it must be considered for the full life cycle of the control. Annual loss expectancy (ALE) and the frequency of incidents could help measure the benefit, but would have more of an indirect relationship as not all incidents may be mitigated by implementing a two-factor authentication system. The approved budget for the project may have no bearing on what the project may actually cost.

S2-80   One way to determine control effectiveness is by determining:

             A.   whether it is preventive, detective or compensatory.
             B.   the capability of providing notification of failure.
             C.   the test results of intended objectives.
             D.   the evaluation and analysis of reliability.

C   Control effectiveness requires a process to verify that the control process worked as intended. Examples such as dual-control or dual-entry bookkeeping provide verification and assurance that the process operated as intended. The type of control is not relevant, and notification of failure is not determinative of control strength. Reliability is not an indication of control strength; weak controls can be highly reliable, even if they are ineffective controls.

S2-81   What does a network vulnerability assessment intend to identify?

             A.   0-day vulnerabilities
             B.   Malicious software and spyware
             C.   Security design flaws
             D.   Misconfiguration and missing updates

D   A network vulnerability assessment intends to identify known vulnerabilities based on common misconfigurations and missing updates. 0-day vulnerabilities by definition are not previously known and therefore are undetectable. Malicious software and spyware are normally addressed through antivirus and antispyware policies. Security design flaws require a deeper level of analysis.

S2-82   Who is responsible for ensuring that information is classified?

             A.   Senior management
             B.   Security manager
             C.   Data owner
             D.   Custodian

C   The data owner is responsible for applying the proper classification to the data. Senior management is ultimately responsible for the organization. The security officer is responsible for applying security protection relative to the level of classification specified by the owner. The technology group is delegated the custody of the data by the data owner, but the group does not classify the information.

S2-83   After a risk assessment, it is determined that the cost to mitigate the risk is much greater than the benefit to be derived. The information security manager should recommend to business management that the risk be:

A.   transferred.
B.   treated.
C.   accepted.
D.   terminated.

C   When the cost of control is more than the cost of the risk, the risk should be accepted. Transferring, treating or terminating the risk is of limited benefit if the cost of that control is more than the cost of the risk itself.

S2-84   The **PRIMARY** reason for initiating a policy exception process is when:

A.   operations are too busy to comply.
B.   the risk is justified by the benefit.
C.   policy compliance would be difficult to enforce.
D.   users may initially be inconvenienced.

B   Exceptions to policy are warranted in circumstances where compliance may be difficult or impossible and the risk of noncompliance is outweighed by the benefits. Being busy is not a justification for policy exceptions, nor is the fact that compliance cannot be enforced. User inconvenience is not a reason to automatically grant exception to a policy.

S2-85   Which of the following would be the **MOST** relevant factor when defining the information classification policy?

A.   Quantity of information
B.   Available IT infrastructure
C.   Benchmarking
D.   Requirements of data owners

D   When defining the information classification policy, the requirements of the data owners need to be identified. The quantity of information, availability of IT infrastructure and benchmarking may be part of the scheme after the fact and would be less relevant.

S2-86   The **MOST** appropriate owner of customer data stored in a central database, used only by an organization's sales department, would be the:

A.   sales department.
B.   database administrator.
C.   chief information officer (CIO).
D.   head of the sales department.

D   The owner of the information asset should be the person with the decision-making power in the department deriving the most benefit from the asset. In this case, it would be the head of the sales department. The organizational unit cannot be the owner of the asset because that removes personal responsibility. The database administrator is a custodian. The chief information officer (CIO) would not be an owner of this database because the CIO is less likely to be knowledgeable about the specific needs of sales operations and security concerns.

S2-87   In assessing the degree to which an organization may be affected by new privacy legislation, information security management should **FIRST**:

    A. develop an operational plan for achieving compliance with the legislation.
    B. identify systems and processes that contain privacy components.
    C. restrict the collection of personal information until compliant.
    D. identify privacy legislation in other countries that may contain similar requirements.

B   Identifying the relevant systems and processes is the best first step. Developing an operational plan for achieving compliance with the legislation is incorrect because it is not the first step. Restricting the collection of personal information comes later. Identifying privacy legislation in other countries would not add much value.

S2-88   Risk assessment is **MOST** effective when performed:

    A. at the beginning of security program development.
    B. on a continuous basis.
    C. while developing the business case for the security program.
    D. during the business change process.

B   Risk assessment needs to be performed on a continuous basis because of organizational and technical changes. Risk assessment must take into account all significant changes in order to be effective.

S2-89   Which of the following is the **MAIN** reason for performing risk assessment on a continuous basis?

    A. Justification of the security budget must be continually made.
    B. New vulnerabilities are discovered every day.
    C. The risk environment is constantly changing.
    D. Management needs to be continually informed about emerging risks.

C   The risk environment is impacted by factors such as changes in technology, and business strategy. These changes introduce new threats and vulnerabilities to the organization. As a result, risk assessment should be performed continuously. Justification of a budget should never be the main reason for performing a risk assessment. New vulnerabilities should be managed through a patch management process. Informing management about emerging risks is important, but is not the main driver for determining when a risk assessment should be performed.

S2-90   There is a time lag between the time when a security vulnerability is first published, and the time when a patch is delivered. Which of the following should be carried out **FIRST** to mitigate the risk during this time period?

    A. Identify the vulnerable systems and apply compensating controls
    B. Minimize the use of vulnerable systems
    C. Communicate the vulnerability to system users
    D. Update the signatures database of the intrusion detection system (IDS)

A   The best protection is to identify the vulnerable systems and apply compensating controls until a patch is installed. Minimizing the use of vulnerable systems and communicating the vulnerability to system users could be compensating controls but would not be the first course of action. Choice D does not make clear the timing of when the intrusion detection system (IDS) signature list would be updated to accommodate the vulnerabilities that are not yet publicly known. Therefore, this approach should not always be considered as the first option.

CISM Review Questions, Answers & Explanations Manual 2012   **65**
ISACA. All Rights Reserved.

S2-91   Which of the following security activities should be implemented in the change management process to identify key vulnerabilities introduced by changes?

        A. Business impact analysis (BIA)
        B. Penetration testing
        C. Audit and review
        D. Threat analysis

B   Penetration testing focuses on identifying vulnerabilities. None of the other choices would identify vulnerabilities introduced by changes.

S2-92   Which of the following techniques **MOST** clearly indicates whether specific risk-reduction controls should be implemented?

        A. Cost-benefit analysis
        B. Penetration testing
        C. Frequent risk assessment programs
        D. Annual loss expectancy (ALE) calculation

A   In a cost-benefit analysis, the annual cost of safeguards is compared with the expected cost of loss. This can then be used to justify a specific control measure. Penetration testing may indicate the extent of a weakness but, by itself, will not establish the cost/benefit of a control. Frequent risk assessment programs will certainly establish what risk exists but will not determine the maximum cost of controls. Annual loss expectancy (ALE) is a measure which will contribute to the value of the risk but, alone, will not justify a control.

S2-93   An organization has decided to implement additional security controls to treat the risks of a new process. This is an example of:

        A. eliminating the risk.
        B. transferring the risk.
        C. mitigating the risk.
        D. accepting the risk.

C   Risk can never be eliminated entirely. Transferring the risk gives it away such as buying insurance so the insurance company can take the risk. Implementing additional controls is an example of mitigating risk. Doing nothing to mitigate the risk would be an example of accepting risk.

S2-94   Which of the following roles is **PRIMARILY** responsible for determining the information classification levels for a given information asset?

        A. Manager
        B. Custodian
        C. User
        D. Owner

D   Although the information owner may be in a management position and is also considered a user, the information owner role has the responsibility for determining information classification levels. Management is responsible for higher-level issues such as providing and approving budget, supporting activities, etc. The information custodian is responsible for day-to-day security tasks such as protecting information, backing up information, etc. Users are the lowest level. They use the data, but do not classify the data. The owner classifies the data.

S2-95   The **PRIMARY** reason for assigning classes of sensitivity and criticality to information resources is to provide a basis for:

A. determining the scope for inclusion in an information security program.
B. defining the level of access controls.
C. justifying costs for information resources.
D. determining the overall budget of an information security program.

B   The assigned class of sensitivity and criticality of the information resource determines the level of access controls to be put in place. The assignment of sensitivity and criticality takes place with the information assets that have already been included in the information security program and has only an indirect bearing on the costs to be incurred. The assignment of sensitivity and criticality contributes to, but does not decide, the overall budget of the information security program.

S2-96   An organization is already certified to an international security standard. Which mechanism would **BEST** help to further align the organization with other data security regulatory requirements as per new business needs?

A. Key performance indicators (KPIs)
B. Business impact analysis (BIA)
C. Gap analysis
D. Technical vulnerability assessment

C   Gap analysis would help identify the actual gaps between the desired state and the current implementation of information security management. BIA is primarily used for business continuity planning. Technical vulnerability assessment is used for detailed assessment of technical controls, which would come later in the process and would not provide complete information in order to identify gaps.

S2-97   When performing a qualitative risk analysis, which of the following will **BEST** produce reliable results?

A. Estimated productivity losses
B. Possible scenarios with threats and impacts
C. Value of information assets
D. Vulnerability assessment

B   Listing all possible scenarios that could occur, along with threats and impacts, will better frame the range of risks and facilitate a more informed discussion and decision. Estimated productivity losses, value of information assets and vulnerability assessments would not be sufficient on their own.

S2-98   Which of the following is the **BEST** method to ensure the overall effectiveness of a risk management program?

A. User assessments of changes
B. Comparison of the program results with industry standards
C. Assignment of risk within the organization
D. Participation by all members of the organization

D   Effective risk management requires participation, support and acceptance by all applicable members of the organization, beginning with the executive levels. Personnel must understand their responsibilities and be trained on how to fulfill their roles.

S2-99   The **MOST** effective use of a risk register is to:

          A. identify risks and assign roles and responsibilities for mitigation.
          B. identify threats and probabilities.
          C. facilitate a thorough review of all IT-related risks on a periodic basis.
          D. record the annualized financial amount of expected losses due to risks.

C   A risk register is more than a simple list--it should be used as a tool to ensure comprehensive
          documentation, periodic review and formal update of all risk elements in the enterprise's IT and related
          organization. Identifying risks and assigning roles and responsibilities for mitigation are elements of
          the register. Identifying threats and probabilities are two elements that are defined in the risk matrix, as
          differentiated from the broader scope of content in, and purpose for, the risk register. While the annualized
          loss expectancy (ALE) should be included in the register, this quantification is only a single element in the
          overall risk analysis program.

S2-100   Which of the following are the essential ingredients of a business impact analysis (BIA)?

          A. Downtime tolerance, resources and criticality
          B. Cost of business outages in a year as a factor of the security budget
          C. Business continuity testing methodology being deployed
          D. Structure of the crisis management team

A   The main purpose of a BIA is to measure the downtime tolerance, associated resources and criticality of
          a business function. Options B, C and D are all associated with business continuity planning, but are not
          related to the BIA.

S2-101   A risk management approach to information protection is:

          A. managing risks to an acceptable level, commensurate with goals and objectives.
          B. accepting the security posture provided by commercial security products.
          C. implementing a training program to educate individuals on information protection and risks.
          D. managing risk tools to ensure that they assess all information protection vulnerabilities.

A   Risk management is identifying all risks within an organization, establishing an acceptable level of risk
          and effectively managing risks which may include mitigation or transfer. Accepting the security posture
          provided by commercial security products is an approach that would be limited to technology components
          and may not address all business operations of the organization. Education is a part of the overall risk
          management process. Tools may be limited to technology and would not address non-technology risks.

S2-102   Which of the following is the **MOST** effective way to treat a risk such as a natural disaster that has a low
          probability and a high impact level?

          A. Implement countermeasures.
          B. Eliminate the risk.
          C. Transfer the risk.
          D. Accept the risk.

C   Risks are typically transferred to insurance companies when the probability of an incident is low but the
          impact is high. Examples include: hurricanes, tornados and earthquakes. Implementing countermeasures
          may not be the most cost-effective approach to security management. Eliminating the risk may not be
          possible. Accepting the risk would leave the organization vulnerable to a catastrophic disaster which may
          cripple or ruin the organization. It would be more cost effective to pay recurring insurance costs than to be
          affected by a disaster from which the organization cannot financially recover.

S2-103 When implementing security controls, an information security manager must **PRIMARILY** focus on:

    A. minimizing operational impacts.
    B. eliminating all vulnerabilities.
    C. usage by similar organizations.
    D. certification from a third party.

A Security controls must be compatible with business needs. It is not feasible to eliminate all vulnerabilities. Usage by similar organizations does not guarantee that controls are adequate. Certification by a third party is important, but not a primary concern.

S2-104 All risk management activities are **PRIMARILY** designed to reduce impacts to:

    A. a level defined by the security manager.
    B. an acceptable level based on organizational risk tolerance.
    C. a minimum level consistent with regulatory requirements.
    D. the minimum level possible.

B The aim of risk management is to reduce impacts to an acceptable level. "Acceptable" or "reasonable" are relative terms that can vary based on environment and circumstances. A minimum level that is consistent with regulatory requirements may not be consistent with business objectives, and regulators typically do not assign risk levels. The minimum level possible may not be aligned with business requirements.

S2-105 Which of the following is the **MOST** important requirement for setting up an information security infrastructure for a new system?

    A. Performing a business impact analysis (BIA)
    B. Considering personal information devices as part of the security policy
    C. Initiating IT security training and familiarization
    D. Basing the information security infrastructure on risk assessment

D The information security infrastructure should be based on risk. While considering personal information devices as part of the security policy may be a consideration, it is not the most important requirement. A BIA is typically carried out to prioritize business processes as part of a business continuity plan. Initiating IT security training may not be important for the purpose of the information security infrastructure.

S2-106 Previously accepted risk should be:

    A. reassessed periodically since the risk can be escalated to an unacceptable level due to changing conditions.
    B. accepted permanently since management has already spent resources (time and labor) to conclude that the risk level is acceptable.
    C. avoided next time since risk avoidance provides the best protection to the company.
    D. removed from the risk log once it is accepted.

A Acceptance of risk should be regularly reviewed to ensure that the rationale for the initial risk acceptance is still valid within the current business context. The rationale for initial risk acceptance may no longer be valid due to change(s) and, hence, risk cannot be accepted permanently. Risk is an inherent part of business and it is impractical and costly to eliminate all risk. Even risks that have been accepted should be monitored for changing conditions that could alter the original decision.

S2-107   An information security manager is advised by contacts in law enforcement that there is evidence that his/her company is being targeted by a skilled gang of hackers known to use a variety of techniques, including social engineering and network penetration. The **FIRST** step that the security manager should take is to:

     A.   perform a comprehensive assessment of the organization's exposure to the hacker's techniques.
     B.   initiate awareness training to counter social engineering.
     C.   immediately advise senior management of the elevated risk.
     D.   increase monitoring activities to provide early detection of intrusion.

C   Information about possible significant new risks from credible sources should be provided to management along with advice on steps that need to be taken to counter the threat. The security manager should assess the risk, but senior management should be immediately advised. It may be prudent to initiate an awareness campaign subsequent to sounding the alarm if awareness training is not current. Monitoring activities should also be increased.

S2-108   Which of the following steps should be performed **FIRST** in the risk assessment process?

     A.   Staff interviews
     B.   Threat identification
     C.   Asset identification and valuation
     D.   Determination of the likelihood of identified risks

C   The first step in the risk assessment methodology is a system characterization, or identification and valuation, of all of the enterprise's assets to define the boundaries of the assessment. Interviewing is a valuable tool to determine qualitative information about an organization's objectives and tolerance for risk. Interviews are used in subsequent steps. Identification of threats comes later in the process and should not be performed prior to an inventory since many possible threats will not be applicable if there is no asset at risk. Determination of likelihood comes later in the risk assessment process.

S2-109   Which of the following authentication methods prevents authentication replay?

     A.   Password hash implementation
     B.   Challenge/response mechanism
     C.   Wired Equivalent Privacy (WEP) encryption usage
     D.   HTTP Basic Authentication

B   A challenge/response mechanism prevents replay attacks by sending a different random challenge in each authentication event. The response is linked to that challenge. Therefore, capturing the authentication handshake and replaying it through the network will not work. Using hashes by itself will not prevent a replay. A WEP key will not prevent sniffing (it just takes a few more minutes to break the WEP key if the attacker does not already have it) and therefore will not be able to prevent recording and replaying an authentication handshake. HTTP Basic Authentication is clear text and has no mechanisms to prevent replay.

S2-110   An organization has a process in place that involves the use of a vendor. A risk assessment was completed during the development of the process. A year after the implementation a monetary decision has been made to use a different vendor. What, if anything, should occur?

        A. Nothing, since a risk assessment was completed during development.
        B. A vulnerability assessment should be conducted.
        C. A new risk assessment should be performed.
        D. The new vendor's SAS 70 type II report should be reviewed.

C   The risk assessment process is continual and any changes to an established process should include a new risk assessment. While a review of the SAS 70 report and a vulnerability assessment may be components of a risk assessment, neither would constitute sufficient due diligence on its own.

S2-111   Which of the following would govern which information assets need more protection than other information assets?

        A. A data custodian
        B. Asset identification
        C. Data classification
        D. An acceptable use policy

C   Classification defines which assets are more critical or sensitive and thus need more protection. A data custodian is not responsible for determining the level of protection, but for ensuring that controls are in place and in compliance with the classification requirements. By mere identification of assets, one cannot be sure which assets are more important than others. An acceptable use policy will only govern the acceptable use of various enterprise assets and will not help in establishing which assets are more important than others.

S2-112   When calculating an annual loss expectancy (ALE), which variable **MOST** requires the information systems (IS) manager to form an opinion based on the uncertainty of the future?

        A. Exposure factor
        B. Asset value
        C. Annual rate of occurrence
        D. Recovery time objective (RTO)

C   The annual rate of occurrence is a factor. The other choices do not require forming an opinion about the ALE.

S2-113   A security risk assessment of a proposed software purchase was conducted and the software was found to have a low level of risk to the enterprise. After the purchase and implementation of the software, an unexpected vulnerability materialized and was subsequently remediated. What should the information security manager do **NEXT**?

A.   Revise the risk assessment process to ensure that all vulnerabilities are identified in future assessments.
B.   Determine whether the product description was misleading and seek compensation from the developer for time and damages.
C.   Return the product and perform a new selection risk assessment to ensure that problems are avoided in the future.
D.   Revisit and update the risk assessment to account for the unforeseen vulnerability.

D   Risk assessments should be updated periodically to assess new vulnerabilities to determine their impact on the business. No risk assessment can ever ensure that all vulnerabilities are identified. Determining whether the product description was misleading and seeking compensation from the developer for time and damages is not necessarily part of the risk assessment process. Returning the product and performing a new selection risk assessment to ensure that problems are avoided in the future is not the next step.

S2-114   The goals of information security risk management inside an enterprise are **BEST** achieved if these risk management activities are:

A.   treated as a distinct process.
B.   conducted by the IT department.
C.   integrated within business processes.
D.   communicated to all employees.

C   Risk management activities are more likely to be executed as part of a business process. The scope of information security risk management encompasses more than IT processes. Communication alone does not necessarily correlate with successful execution of the process.

S2-115   The **PRIMARY** objective when selecting controls and countermeasures is to:

A.   protect against all threats.
B.   reduce costs.
C.   optimize protection and usability.
D.   restrict employee access.

C   Optimized controls are understood to be cost-effective and should provide the appropriate level of protection. It is not feasible to protect against all threats. Business needs could require more expensive controls. Restriction of employee access may be part of a control, but it is not the objective of a control.

S2-116   Segregation of duties assists with:

A.   employee monitoring.
B.   reduced supervisory requirements.
C.   fraud prevention.
D.   enhanced compliance.

C   Segregation of duties is primarily used to discourage fraudulent activities. Employee monitoring and enhanced compliance are secondary considerations. Supervision is not reduced, but facilitated.

S2-117   The selection and implementation of products to protect the IT infrastructure should be **PRIMARILY** based on:

    A. regulatory requirements.
    B. a technical expert advisory.
    C. state-of-the-art technology.
    D. a risk assessment.

D   A risk assessment helps identify control gaps in the IT infrastructure and prioritize mitigation plans, which will help drive selection of security solutions. Regulatory requirements drive business requirements. An expert advisory may not be aligned with business needs. A risk assessment is the main driver for selecting technologies.

S2-118   The **BEST** way to standardize security configurations in similar devices is through the use of:

    A. policies.
    B. procedures.
    C. technical guides.
    D. baselines.

D   Baselines describe the minimum configuration requirements across similar devices, activities or resources. Policies set high-level direction, not technical details. Procedures are use to provide instructions, not configuration details. Technical guides provide support, but not necessarily the requirements.

S2-119   Which of the following, if not properly secured and/or segregated from local area network (LAN), can be the **GREATEST** source of risk to an enterprise's internal network?

    A. A virtual local area network (VLAN)
    B. A virtual private network (VPN)
    C. A wireless local area network (WLAN)
    D. Voice-over IP (VoIP)

C   If not properly configured, a WLAN is not secure and can expose the network to unauthorized external access. Even if misconfigured, the other choices generally pose less risk.

S2-120   Which of the following would provide the **BEST** defense against the introduction of malware in end-user computers via the Internet browser?

    A. Input validation checks on SQL injection
    B. Restricting access to social media sites
    C. Deleting temporary files
    D. Restricting execution of mobile code

D   Restricting execution of mobile code is the most effective way to avoid introduction of malware into the end user's computers. Validation of checks on SQL injection does not apply to this scenario. Restricting access to social media sites may be helpful, but is not the primary source of malware. Deleting temporary files is not applicable to this scenario.

S2-121   An enterprise is transferring its IT operations to an offshore location. An information security manager should be **PRIMARILY** concerned about:

A. reviewing new laws and regulations.
B. updating operational procedures.
C. validating staff qualifications.
D. conducting a risk assessment.

D   A risk assessment should be conducted to determine new risks introduced by the outsourced processes. The other choices may or may not be identified as mitigating measures based on the risks determined by the assessment.

S2-122   Which of the following statements concerning the transfer of risk is **TRUE**?

A. Responsibility cannot be transferred.
B. Transferring risk is a form of mitigation.
C. Transferring risk eliminates the risk.
D. Risk cannot be transferred.

A   Transferring risk is a compensatory control that serves to reduce impact, but does not eliminate responsibility. Transfer of risk does not mitigate the risk, but is a parallel option to mitigation (reduction). Transfer does not eliminate the risk. Some risk cannot be transferred.

S2-123   An information security manager is working with a business manager to develop risk management strategies for an application. The business manager believes that using an external application service provider (ASP) will eliminate all of its risk. Which of the following should the security manager explain to the business manager?

A. Outsourcing will only transfer some of the risk.
B. Additional insurance must be purchased to completely cover the risk.
C. All four types of risk treatment strategies must be applied to ensure that all risks are eliminated.
D. Outsourcing will not mitigate any risk.

A   Outsourcing is a form of risk transfer, but will not eliminate all risks to the enterprise and will not absolve the enterprise of responsibility for those risks. When an enterprise uses an outsourcing provider, some of the technical and financial risks are transferred to the outsourcing organization at a cost that should reflect financial or other benefits to the enterprise. Responsibility for the risk is actually not transferred; the potential impact to the enterprise is reduced to the extent that the benefits of outsourcing will outweigh the risks associated with a compromise. Additional insurance may or may not be necessary since proper insurance may already be included in the contract with the outsource provider. Additionally, insurance does not eliminate risk; it is used only to reduce the financial impact of loss due to an insured event. While all strategies must be considered and analyzed, the most cost-effective solution may not include all strategies. Additionally, it is not possible to eliminate all risks. Outsourcing does mitigate some risk.

S2-124   The use of insurance is an example of which of the following?

A. Risk mitigation
B. Risk acceptance
C. Risk elimination
D. Risk transfer

D   Insurance is a method of offsetting the financial loss that might be incurred as a result of an adverse event. Some, but not all, of the potential costs are transferred to the insurance company. The effects from a potential event can be shared by procuring assurance, but risks themselves are not mitigated. Acceptance of risk is a decision by the enterprise to assume the impact of the effects of an event. Risks are never fully eliminated, unless the activity that causes the risk is stopped or avoided.

S2-125   Once residual risks have been determined, the enterprise should **NEXT**:

A. transfer the remaining risks to a third party.
B. acquire insurance against the effects of the residual risks.
C. validate that the residual risks are acceptable.
D. formally document and accept the residual risks.

C   Once residual risks have been determined, the next step should be to validate that they are acceptable and within the enterprise's risk tolerance. All other options are steps taken after initial validation occurs.

S2-126   An information security manager performing a security review determines that compliance with access control policies to the data center is inconsistent across employees. The **FIRST** step to address this issue should be to:

A. assess the risk of noncompliance.
B. initiate security awareness training.
C. prepare a status report for management.
D. increase compliance enforcement.

A   Inconsistent compliance can be the result of different factors, but is often a lack of awareness. Assessing the risk of noncompliance will provide the information needed to determine the most effective remediation requirements. If awareness is adequate, training may not help and increased compliance enforcement may be indicated. A report may be warranted, but will not directly address the issue that is normally a part of the information security manager's responsibilities. Increased enforcement is not warranted if the problem is a lack of effective communication about security policy.

S2-127   Which of the following is the **MOST** important element to consider when initiating asset classification?

A. The type of IT hardware that must be classified
B. A comprehensive risk assessment and analysis
C. Business continuity and disaster recovery plans (BCPs/DRPs)
D. The consequences of losing system functionality

D   Business criticality and sensitivity is the primary consideration for a classification scheme. This is determined by a business impact analysis (BIA), which will determine the consequences of losing or compromising various information systems. The type of hardware is typically not a classification issue, although it is a factor in incident recovery considerations. Classification is concerned with the loss or compromise of information systems, not the risk they are subject to. Classification is an element of BCP/DRP, but is not required for classification.

S2-128   Which of the following factors will **MOST** affect the extent to which controls should be layered?

A.   The extent to which controls are procedural
B.   Controls subject to the same threat
C.   The maintenance cost of controls
D.   Controls that fail in a closed condition

B   To manage the aggregate risk of total risk, common failure modes in existing controls must be addressed by adding or modifying controls so that they fail under different conditions. Whether controls are procedural or technical will not affect layering requirements. Excessive maintenance costs will probably not be addressed by adding additional controls. Controls that fail in a closed condition pose risks to availability, whereas controls that fail in an open condition may require additional control layers to prevent compromise.

S2-129   Generally, who should determine the classification of an information asset?

A.   The asset custodian
B.   The security manager
C.   Senior management
D.   The asset owner

D   Classifying an information asset is the responsibility of the asset owner.

S2-130   Which of the following is a preventive measure?

A.   A warning banner
B.   Audit trails
C.   An access control
D.   An alarm system

C   Preventive controls inhibit attempts to violate security policies. An example of such a control is an access control. A warning banner is a deterrent control, which provides a warning that can deter potential compromise. Audit trails are an example of a detective control. An alarm system is an example of a detective control.

S2-131   When a security standard conflicts with a business objective, the situation should be resolved by:

A.   changing the security standard.
B.   changing the business objective.
C.   performing a risk analysis.
D.   authorizing a risk acceptance.

C   Conflicts of this type should be based on a risk analysis of the costs and benefits of allowing or disallowing an exception to the standard. It is highly improbable that a business objective could be changed to accommodate a security standard, while risk acceptance is a process that derives from the risk analysis.

S2-132   A business unit intends to deploy a new technology in a manner that places it in violation of existing information security standards. What immediate action should an information security manager take?

A. Enforce the existing security standard
B. Change the standard to permit the deployment
C. Perform a risk analysis to quantify the risk
D. Perform research to propose use of a better technology

C   Resolving conflicts of this type should be based on a sound risk analysis of the costs and benefits of allowing or disallowing an exception to the standard. A blanket decision should never be given without conducting such an analysis. Enforcing existing standards is a good practice; however, standards need to be continuously examined in light of new technologies and the risks they present. Standards should not be changed without an appropriate risk assessment.

S2-133   Logging is an example of which type of defense against systems compromise?

A. Containment
B. Detection
C. Reaction
D. Recovery

B   Detection defenses include logging as well as monitoring, measuring, auditing, detecting viruses and intrusion. Examples of containment defenses are awareness, training and physical security defenses. Examples of reaction defenses are incident response, policy and procedure change, and control enhancement. Examples of recovery defenses are backups and restorations, failover and remote sites, and business continuity plans and disaster recovery plans.

S2-134   Temporarily deactivating some monitoring processes, even if supported by an acceptance of operational risk, may not be acceptable to the information security manager if:

A. it implies compliance risks.
B. short-term impact cannot be determined.
C. it violates industry security practices.
D. changes in the roles matrix cannot be detected.

A   Monitoring processes are also required to guarantee fulfillment of laws and regulations of the organization and, therefore, the information security manager will be obligated to comply with the law. Choices B and C are evaluated as part of the operational risk. Choice D is unlikely to be as critical a breach of regulatory legislation. The acceptance of operational risks overrides choices B, C and D.

S2-135   Which of the following is the **MOST** important to keep in mind when assessing the value of information?

A. The potential financial loss
B. The cost of recreating the information
C. The cost of insurance coverage
D. Regulatory requirement

A   The potential for financial loss is always a key factor when assessing the value of information. Choices B, C and D may be contributors, but not the key factor.

S2-136   When a proposed system change violates an existing security standard, the conflict would be **BEST** resolved by:

A.   calculating the risk.
B.   enforcing the security standard.
C.   redesigning the system change.
D.   implementing mitigating controls.

A   Decisions regarding security should always weigh the potential loss from a risk against the existing controls. Each situation is unique; therefore, it is not advisable to always decide in favor of enforcing a standard. Redesigning the proposed change might not always be the best option because it might not meet the business needs. Implementing additional controls might be an option, but this would be done after the risk is known.

S2-137   The information classification scheme should:

A.   consider possible impact of a security breach.
B.   classify personal information in electronic form.
C.   be performed by the information security manager.
D.   classify systems according to the data processed.

A   Data classification is determined by the business risk, i.e., the potential impact on the business of the loss, corruption or disclosure of information. It must be applied to information in all forms, both electronic and physical (paper), and should be applied by the data owner, not the security manager. Choice B is an incomplete answer because it addresses only privacy issues, while choice A is a more complete response. Systems are not classified per se, but the data they process and store should definitely be classified.

S2-138   Which of the following is the **BEST** method to provide a new user with their initial password for e-mail system access?

A.   Interoffice a system-generated complex password with 30 days expiration
B.   Provide a temporary password over the telephone set for immediate expiration
C.   Require no password but force the user to set their own in 10 days
D.   Set initial password equal to the user ID with expiration in 30 days

B   Documenting the password on paper is not the best method even if sent through interoffice mail—if the password is complex and difficult to memorize, the user will likely keep the printed password and this creates a security concern. A temporary password that will need to be changed upon first logon is the best method because it is reset immediately and replaced with the user's choice of password, which will make it easier for the user to remember. If it is given to the wrong person, the legitimate user will likely notify security if still unable to access the system, so the security risk is low. Setting an account with no initial password is a security concern even if it is just for a few days. Choice D provides the greatest security threat because user IDs are typically known by both users and security staff, thus compromising access for up to 30 days.

S2-139   An operating system (OS) noncritical patch to enhance system security cannot be applied because a critical application is not compatible with the change. Which of the following is the **BEST** solution?

    A.   Rewrite the application to conform to the upgraded operating system
    B.   Compensate for not installing the patch with mitigating controls
    C.   Alter the patch to allow the application to run in a privileged state
    D.   Run the application on a test platform; tune production to allow patch and application

B   Since the operating system (OS) patch will adversely impact a critical application, a mitigating control should be identified that will provide an equivalent level of security. Since the application is critical, the patch should not be applied without regard for the application; business requirements must be considered. Altering the OS patch to allow the application to run in a privileged state may create new security weaknesses. Finally, running a production application on a test platform is not an acceptable alternative since it will mean running a critical production application on a platform not subject to the same level of security controls.

S2-140   Primary direction on the impact of compliance with new regulatory requirements that may lead to major application system changes should be obtained from the:

    A.   corporate internal auditor.
    B.   system developers/analysts.
    C.   key business process owners.
    D.   corporate legal counsel.

C   Business process owners are in the best position to understand how new regulatory requirements may affect their systems. Legal counsel and infrastructure management, as well as internal auditors, would not be in as good a position to fully understand all ramifications.

S2-141   The IT function has declared that, when putting a new application into production, it is not necessary to update the business impact analysis (BIA) because it does not produce modifications in the business processes. The information security manager should:

    A.   verify the decision with the business units.
    B.   check the system's risk analysis.
    C.   recommend update after postimplementation review.
    D.   request an audit review.

A   Verifying the decision with the business units is the correct answer because it is not the IT function's responsibility to decide whether a new application modifies business processes Choice B does not consider the change in the applications. Choices C and D delay the update.

S2-142   An internal review of a web-based application system finds the ability to gain access to all employees'
          accounts by changing the employee's ID on the URL used for accessing the account. The vulnerability
          identified is:

          A.   broken authentication.
          B.   unvalidated input.
          C.   cross-site scripting.
          D.   structured query language (SQL) injection.

A   The authentication process is broken because, although the session is valid, the application should
          reauthenticate when the input parameters are changed. The review provided valid employee IDs, and
          valid input was processed. The problem here is the lack of reauthentication when the input parameters are
          changed. Cross-site scripting is not the problem in this case since the attack is not transferred to any other
          user's browser to obtain the output. Structured query language (SQL) injection is not a problem since input
          is provided as a valid employee ID and no SQL queries are injected to provide the output.

S2-143   What is the **MOST** cost-effective method of identifying new vendor vulnerabilities?

          A.   External vulnerability reporting sources
          B.   Periodic vulnerability assessments performed by consultants
          C.   Intrusion prevention software
          D.   Honeypots located in the DMZ

A   External vulnerability sources are going to be the most cost-effective method of identifying these vulnerabilities.
          The cost involved in choices B and C would be much higher, especially if performed at regular intervals.
          Honeypots would not identify all vendor vulnerabilities. In addition, honeypots located in the DMZ can create a
          security risk if the production network is not well protected from traffic from compromised honeypots.

S2-144   Of the following, retention of business records should be **PRIMARILY** based on:

          A.   periodic vulnerability assessment.
          B.   regulatory and legal requirements.
          C.   device storage capacity and longevity.
          D.   past litigation.

B   Retention of business records is a business requirement that must consider regulatory and legal
          requirements based on geographic location and industry. Choices A and C are important elements for
          making the decision, but the primary driver is the legal and regulatory requirements that need to be
          followed by all companies. Record retention may take into consideration past litigation, but it should not be
          the primary decision factor.

S2-145   Which is the **BEST** way to measure and prioritize aggregate risk deriving from a chain of linked system vulnerabilities?

          A. Vulnerability scans
          B. Penetration tests
          C. Code reviews
          D. Security audits

B   A penetration test is normally the only security assessment that can link vulnerabilities together by exploiting them sequentially. This gives a good measurement and prioritization of risks. Other security assessments such as vulnerability scans, code reviews and security audits can help give an extensive and thorough risk and vulnerability overview, but will not be able to test or demonstrate the final consequence of having several vulnerabilities linked together. Penetration testing can give risk a new perspective and prioritize based on the end result of a sequence of security problems.

S2-146   Determining the nature and extent of activities required in developing or improving an information security program often requires assessing the existing security levels of various program components. The **BEST** process to accomplish this task is to perform a(n):

          A. impact assessment.
          B. vulnerability assessment.
          C. gap analysis.
          D. threat assessment.

C   A gap analysis is used to determine the current state of security for various program components as compared to the desired state. Once the gaps have been determined, action items to improve various aspects of the program should be prioritized using a risk-based approach. An impact assessment is used to determine potential impact in the event of loss of a resource. Vulnerability is only one aspect to be considered in a security review. A threat assessment would not normally be a part of a security review.

S2-147   The design and implementation of controls and countermeasures must be **PRIMARILY** focused on:

          A. eliminating IT risk.
          B. cost-benefit balance.
          C. resource management.
          D. the number of assets protected.

B   The balance between cost and benefits should direct controls selection. The focus must include procedural, operational and other risks, in addition to IT risk. Resource management is not directly related to controls. The implementation of controls is based on the impact and risk, not on the number of assets.

S2-148   The **PRIMARY** purpose of performing an internal attack and penetration test is to identify:

          A. weaknesses in network and server security.
          B. ways to improve the incident response process.
          C. potential attack vectors on the network perimeter.
          D. the optimum response to internal hacker attacks.

A   An internal attack and penetration test are designed to identify weaknesses in network and server security. They do not focus as much on incident response or the network perimeter.

S2-149 An organization has learned of a security breach at another company that utilizes similar technology. The **FIRST** thing the information security manager should do is:

A. assess the likelihood of incidents from the reported cause.
B. discontinue the use of the vulnerable technology.
C. report to senior management that the organization is not affected.
D. remind staff that no similar security breaches have taken place.

A The security manager should first assess the likelihood of a similar incident occurring, based on available information. Discontinuing the use of the vulnerable technology would not necessarily be practical since it would likely be needed to support the business. Reporting to senior management that the organization is not affected due to controls already in place would be premature until the information security manager can first assess the impact of the incident. Until this has been researched, it is not certain that no similar security breaches have taken place.

DOMAIN 3—INFORMATION SECURITY PROGRAM DEVELOPMENT AND MANAGEMENT (25%)

S3-1   Who can **BEST** advocate the development of and ensure the success of an information security program?

A. Internal auditor
B. Chief operating officer (COO)
C. Steering committee
D. IT management

C   Senior management represented in the security steering committee is in the best position to advocate the establishment of and continued support for an information security program. The chief operating officer (COO) will be a member of that committee. An internal auditor is a good advocate but is secondary to the influence of senior management. IT management has a lesser degree of influence and would also be part of the steering committee.

S3-2   Which of the following **BEST** ensures that information transmitted over the Internet will remain confidential?

A. Virtual private network (VPN)
B. Firewalls and routers
C. Biometric authentication
D. Two-factor authentication

A   Encryption of data in a virtual private network (VPN) ensures that transmitted information is not readable, even if intercepted. Firewalls and routers protect access to data resources inside the network and do not protect traffic in the public network. Biometric and two-factor authentication, by themselves, would not prevent a message from being intercepted and read.

S3-3   The effectiveness of virus detection software is **MOST** dependent on which of the following?

A. Packet filtering
B. Intrusion detection
C. Software upgrades
D. Definition files

D   The effectiveness of virus detection software depends on virus signatures which are stored in virus definition files. Software upgrades are related to the periodic updating of the program code, which would not be as critical. Intrusion detection and packet filtering do not focus on virus detection.

S3-4   Which of the following is the **MOST** effective type of access control?

A. Centralized
B. Role-based
C. Decentralized
D. Discretionary

B   Role-based access control allows users to be grouped into job-related categories, which significantly eases the required administrative overhead. Discretionary access control would require a greater degree of administrative overhead. Decentralized access control generally requires a greater number of staff to administer, while centralized access control is an incomplete answer.

S3-5   Which of the following devices should be placed within a DMZ?

     A. Router
     B. Firewall
     C. Mail relay
     D. Authentication server

C   A mail relay should normally be placed within a demilitarized zone (DMZ) to shield the internal network. An authentication server, due to its sensitivity, should always be placed on the internal network, never on a DMZ that is subject to compromise. Both routers and firewalls may bridge a DMZ to another network, but do not technically reside within the DMZ network segment.

S3-6   An intrusion detection system should be placed:

     A. outside the firewall.
     B. on the firewall server.
     C. on a screened subnet.
     D. on the external router.

C   An intrusion detection system (IDS) should be placed on a screened subnet, which is a demilitarized zone (DMZ). Placing it on the Internet side of the firewall is not advised because the system will generate alerts on all malicious traffic—even though 99 percent will be stopped by the firewall and never reach the internal network. The same would be true of placing it on the external router, if such a thing were feasible. Since firewalls should be installed on hardened servers with minimal services enabled, it would be inappropriate to install the IDS on the same physical device.

S3-7   The **BEST** reason for an organization to have two discrete firewalls connected directly to the Internet and to the same DMZ would be to:

     A. provide in-depth defense.
     B. separate test and production.
     C. permit traffic load balancing.
     D. prevent a denial-of-service attack.

C   Having two entry points, each guarded by a separate firewall, is desirable to permit traffic load balancing. As they both connect to the Internet and to the same demilitarized zone (DMZ), such an arrangement is not practical for separating test from production or preventing a denial-of-service attack.

S3-8   An extranet server should be placed:

     A. outside the firewall.
     B. on the firewall server.
     C. on a screened subnet.
     D. on the external router.

C   An extranet server should be placed on a screened subnet, which is a demilitarized zone (DMZ). Placing it on the Internet side of the firewall would leave it defenseless. The same would be true of placing it on the external router, although this would not be possible. Since firewalls should be installed on hardened servers with minimal services enabled, it would be inappropriate to store the extranet on the same physical device.

S3-9   Which of the following is the **BEST** metric for evaluating the effectiveness of security awareness training? The number of:

A. password resets.
B. reported incidents.
C. incidents resolved.
D. access rule violations.

B   Reported incidents will provide an indicator of the awareness level of staff. An increase in reported incidents could indicate that the staff is paying more attention to security. Password resets and access rule violations may or may not have anything to do with awareness levels. The number of incidents resolved may not correlate to staff awareness.

S3-10   Security monitoring mechanisms should **PRIMARILY**:

A. focus on business-critical information.
B. assist owners to manage control risks.
C. focus on detecting network intrusions.
D. record all security violations.

A   Security monitoring must focus on business-critical information to remain effectively usable by and credible to business users. Control risk is the possibility that controls would not detect an incident or error condition, and therefore is not a correct answer because monitoring would not directly assist in managing this risk. Network intrusions are not the only focus of monitoring mechanisms; although they should record all security violations, this is not the primary objective.

S3-11   When contracting with an outsourcer to provide security administration, the **MOST** important contractual element is the:

A. right-to-terminate clause.
B. limitations of liability.
C. service level agreement (SLA).
D. financial penalties clause.

C   Service level agreements (SLAs) provide metrics to which outsourcing firms can be held accountable. This is more important than a limitation on the outsourcing firm's liability, a right-to-terminate clause or a hold-harmless agreement which involves liabilities to third parties.

S3-12   Which of the following is the **BEST** metric for evaluating the effectiveness of an intrusion detection mechanism?

A. Number of attacks detected
B. Number of successful attacks
C. Ratio of false positives to false negatives
D. Ratio of successful to unsuccessful attacks

C   The ratio of false positives to false negatives will indicate whether an intrusion detection system (IDS) is properly tuned to minimize the number of false alarms while, at the same time, minimizing the number of omissions. The number of attacks detected, successful attacks or the ratio of successful to unsuccessful attacks would not indicate whether the IDS is properly configured.

S3-13   Which of the following is **MOST** effective in preventing weaknesses from being introduced into existing production systems?

    A. Patch management
    B. Change management
    C. Security baselines
    D. Virus detection

B   Change management controls the process of introducing changes to systems. This is often the point at which a weakness will be introduced. Patch management involves the correction of software weaknesses and would necessarily follow change management procedures. Security baselines provide minimum recommended settings and do not prevent introduction of control weaknesses. Virus detection is an effective tool but primarily focuses on malicious code from external sources, and only for those applications that are online.

S3-14   Which of the following is **MOST** effective in preventing security weaknesses in operating systems?

    A. Patch management
    B. Change management
    C. Security baselines
    D. Configuration management

A   Patch management corrects discovered weaknesses by applying a correction (a patch) to the original program code. Change management controls the process of introducing changes to systems. Security baselines provide minimum recommended settings. Configuration management controls the updates to the production environment.

S3-15   Which of the following is the **MOST** effective solution for preventing internal users from modifying sensitive and classified information?

    A. Baseline security standards
    B. System access violation logs
    C. Role-based access controls
    D. Exit routines

C   Role-based access controls help ensure that users only have access to files and systems appropriate for their job role. Violation logs are detective and do not prevent unauthorized access. Baseline security standards do not prevent unauthorized access. Exit routines are dependent upon appropriate role-based access.

S3-16   Which of the following is generally used to ensure that information transmitted over the Internet is authentic and actually transmitted by the named sender?

    A. Biometric authentication
    B. Embedded steganographic
    C. Two-factor authentication
    D. Embedded digital signature

D   Digital signatures ensure that transmitted information can be attributed to the named sender; this provides nonrepudiation. Steganographic techniques are used to hide messages or data within other files. Biometric and two-factor authentication is not generally used to protect internet data transmissions.

S3-17   Which of the following is the **MOST** appropriate frequency for updating antivirus signature files for antivirus software on production servers?

    A. Daily
    B. Weekly
    C. Concurrently with O/S patch updates
    D. During scheduled change control updates

A   New viruses are being introduced almost daily. The effectiveness of virus detection software depends on frequent updates to its virus signatures, which are stored on antivirus signature files so updates may be carried out several times during the day. At a minimum, daily updating should occur. Patches may occur less frequently. Weekly updates may potentially allow new viruses to infect the system.

S3-18   Which of the following devices should be placed within a demilitarized zone (DMZ)?

    A. Network switch
    B. Web server
    C. Database server
    D. File/print server

B   A web server should normally be placed within a demilitarized zone (DMZ) to shield the internal network. Database and file/print servers may contain confidential or valuable data and should always be placed on the internal network, never on a DMZ that is subject to compromise. Switches may bridge a DMZ to another network but do not technically reside within the DMZ network segment.

S3-19   On which of the following should a firewall be placed?

    A. Web server
    B. Intrusion detection system (IDS) server
    C. Screened subnet
    D. Domain boundary

D   A firewall should be placed on a (security) domain boundary. Placing it on a web server or screened subnet, which is a demilitarized zone (DMZ), does not provide any protection. Since firewalls should be installed on hardened servers with minimal services enabled, it is inappropriate to have the firewall and the intrusion detection system (IDS) on the same physical device.

S3-20   An intranet server should generally be placed on the:

    A. internal network.
    B. firewall server.
    C. external router.
    D. primary domain controller.

A   An intranet server should be placed on the internal network. Placing it on an external router leaves it defenseless. Since firewalls should be installed on hardened servers with minimal services enabled, it is inappropriate to store the intranet server on the same physical device as the firewall. Similarly, primary domain controllers do not normally share the physical device as the intranet server.

S3-21   Access control to a sensitive intranet application by mobile users can **BEST** be implemented through:

A. data encryption.
B. digital signatures.
C. strong passwords.
D. two-factor authentication.

D   Two-factor authentication through the use of strong passwords combined with security tokens provides the highest level of security. Data encryption, digital signatures and strong passwords do not provide the same level of protection.

S3-22   Security awareness training is **MOST** likely to lead to which of the following?

A. Decrease in intrusion incidents
B. Increase in reported incidents
C. Decrease in security policy changes
D. Increase in access rule violations

B   Reported incidents will provide an indicator as to the awareness level of staff. An increase in reported incidents could indicate that staff is paying more attention to security. Intrusion incidents and access rule violations may or may not have anything to do with awareness levels. A decrease in changes to security policies may or may not correlate to security awareness training.

S3-23   Which of the following ensures that newly identified security weaknesses in an operating system are mitigated in a timely fashion?

A. Patch management
B. Change management
C. Security baselines
D. Acquisition management

A   Patch management involves the correction of software weaknesses and helps ensure that newly identified exploits are mitigated in a timely fashion. Change management controls the process of introducing changes to systems. Security baselines provide minimum recommended settings. Acquisition management controls the purchasing process.

S3-24   The **MAIN** advantage of implementing automated password synchronization is that it:

A. reduces overall administrative workload.
B. increases security between multi-tier systems.
C. allows passwords to be changed less frequently.
D. reduces the need for two-factor authentication.

A   Automated password synchronization reduces the overall administrative workload of resetting passwords. It does not increase security between multi-tier systems, allow passwords to be changed less frequently or reduce the need for two-factor authentication.

S3-25   Which of the following tools is **MOST** appropriate to assess whether information security governance objectives are being met?

    A. SWOT analysis
    B. Waterfall chart
    C. Gap analysis
    D. Balanced scorecard

D   The balanced scorecard is most effective for evaluating the degree to which information security objectives are being met. A SWOT analysis addresses strengths, weaknesses, opportunities and threats. Although useful, a SWOT analysis is not as effective a tool. Similarly, a gap analysis, while useful for identifying the difference between the current state and the desired future state, is not the most appropriate tool. A waterfall chart is used to understand the flow of one process into another.

S3-26   Which of the following is **MOST** effective in preventing the introduction of a code modification that may reduce the security of a critical business application?

    A. Patch management
    B. Change management
    C. Security metrics
    D. Version control

B   Change management controls the process of introducing changes to systems. Failure to have good change management may introduce new weaknesses into otherwise secure systems. Patch management corrects discovered weaknesses by applying a correction to the original program code. Security metrics provide a means for measuring effectiveness. Version control is a subset of change management.

S3-27   Which of the following is **MOST** important to the success of an information security program?

    A. Security awareness training
    B. Achievable goals and objectives
    C. Senior management sponsorship
    D. Adequate start-up budget and staffing

C   Sufficient senior management support is the most important factor for the success of an information security program. Security awareness training, although important, is secondary. Achievable goals and objectives as well as having adequate budgeting and staffing are important factors, but they will not ensure success if senior management support is not present.

S3-28   Which of the following is the **MOST** effective solution for preventing individuals external to the organization from modifying sensitive information on a corporate database?

    A. Screened subnets
    B. Information classification policies and procedures
    C. Role-based access controls
    D. Intrusion detection system (IDS)

A   Screened subnets are demilitarized zones (DMZs) and are oriented toward preventing attacks on an internal network by external users. The policies and procedures to classify information will ultimately result in better protection but they will not prevent actual modification. Role-based access controls would help ensure that users only had access to files and systems appropriate for their job role. Intrusion detection systems (IDS) are useful to detect invalid attempts but they will not prevent attempts.

S3-29   Which of the following technologies is utilized to ensure that an individual connecting to a corporate internal network over the Internet is not an intruder masquerading as an authorized user?

    A. Intrusion detection system (IDS)
    B. IP address packet filtering
    C. Two-factor authentication
    D. Embedded digital signature

C   Two-factor authentication provides an additional security mechanism over and above that provided by passwords alone. This is frequently used by mobile users needing to establish connectivity to a corporate network. IP address packet filtering would protect against spoofing an internal address but would not provide strong authentication. An intrusion detection system (IDS) can be used to detect an external attack but would not help in authenticating a user attempting to connect. Digital signatures ensure that transmitted information can be attributed to the named sender.

S3-30   What is an appropriate frequency for updating operating system (OS) patches on production servers?

    A. During scheduled rollouts of new applications
    B. According to a fixed security patch management schedule
    C. Concurrently with quarterly hardware maintenance
    D. Whenever important security patches are released

D   Patches should be applied whenever important security updates are released. They should not be delayed to coincide with other scheduled rollouts or maintenance. Due to the possibility of creating a system outage, they should not be deployed during critical periods of application activity such as month-end or quarter-end closing.

S3-31   A border router should be placed on which of the following?

    A. Web server
    B. IDS server
    C. Screened subnet
    D. Domain boundary

D   A border router should be placed on a (security) domain boundary. Placing it on a web server or screened subnet, which is a demilitarized zone (DMZ) would not provide any protection. Border routers are positioned on the boundary of the network, but do not reside on a server.

S3-32   An e-commerce order fulfillment web server should generally be placed on which of the following?

    A. Internal network
    B. Demilitarized zone (DMZ)
    C. Database server
    D. Domain controller

B   An e-commerce order fulfillment web server should be placed within a DMZ to protect it and the internal network from external attack. Placing it on the internal network would expose the internal network to potential attack from the Internet. Since a database server should reside on the internal network, the same exposure would exist. Domain controllers would not normally share the same physical device as a web server.

S3-33   Secure customer use of an e-commerce application can **BEST** be accomplished through:

    A. data encryption.
    B. digital signatures.
    C. strong passwords.
    D. two-factor authentication.

A   Encryption would be the preferred method of ensuring confidentiality in customer communications with an e-commerce application. Strong passwords, by themselves, would not be sufficient since the data could still be intercepted, while two-factor authentication would be impractical. Digital signatures would not provide a secure means of communication. In most business-to-customer (B-to-C) web applications, a digital signature is also not a practical solution.

S3-34   What is the **BEST** defense against a Structured Query Language (SQL) injection attack?

    A. Regularly updated signature files
    B. A properly configured firewall
    C. An intrusion detection system
    D. Strict controls on input fields

D   Structured Query Language (SQL) injection involves the typing of programming command statements within a data entry field on a web page, usually with the intent of fooling the application into thinking that a valid password has been entered in the password entry field. The best defense against such an attack is to have strict edits on what can be typed into a data input field so that programming commands will be rejected. Code reviews should also be conducted to ensure that such edits are in place and that there are no inherent weaknesses in the way the code is written; software is available to test for such weaknesses. All other choices would fail to prevent such an attack.

S3-35   Which of the following is the **MOST** important consideration when implementing an intrusion detection system (IDS)?

    A. Tuning
    B. Patching
    C. Encryption
    D. Packet filtering

A   If an intrusion detection system (IDS) is not properly tuned it will generate an unacceptable number of false positives and/or fail to sound an alarm when an actual attack is underway. Patching is more related to operating system hardening, while encryption and packet filtering would not be as relevant.

S3-36   Which of the following is the **MOST** important consideration when securing customer credit card data acquired by a point-of-sale (POS) cash register?

    A. Authentication
    B. Hardening
    C. Encryption
    D. Nonrepudiation

C   Cardholder data should be encrypted using strong encryption techniques. Hardening would be secondary in importance, while nonrepudiation would not be as relevant. Authentication of the point-of-sale (POS) terminal is a previous step to acquiring the card information.

S3-37   Which of the following practices is **BEST** to remove system access for contractors and other temporary users when it is no longer required?

   A.   Log all account usage and send it to their manager
   B.   Establish predetermined automatic expiration dates
   C.   Require managers to e-mail security when the user leaves
   D.   Ensure each individual has signed a security acknowledgement

B   Predetermined expiration dates are the most effective means of removing systems access for temporary users. Reliance on managers to promptly send in termination notices cannot always be counted on, while requiring each individual to sign a security acknowledgement would have little effect in this case.

S3-38   Which of the following is **MOST** important for a successful information security program?

   A.   Adequate training on emerging security technologies
   B.   Open communication with key process owners
   C.   Adequate policies, standards and procedures
   D.   Executive management commitment

D   Sufficient executive management support is the most important factor for the success of an information security program. Open communication, adequate training, and good policies and procedures, while important, are not as important as support from top management; they will not ensure success if senior management support is not present.

S3-39   Which of the following is the **MOST** important item to consider when evaluating products to monitor security across the enterprise?

   A.   Ease of installation
   B.   Product documentation
   C.   Available support
   D.   System overhead

D   Monitoring products can impose a significant impact on system overhead for servers and networks. Product documentation, telephone support and ease of installation, while all important, would be secondary.

S3-40   Which of the following is the **MOST** important guideline when using software to scan for security exposures within a corporate network?

   A.   Never use open source tools
   B.   Focus only on production servers
   C.   Follow a linear process for attacks
   D.   Do not interrupt production processes

D   The first rule of scanning for security exposures is to not break anything. This includes the interruption of any running processes. Open source tools are an excellent resource for performing scans. Scans should focus on both the test and production environments since, if compromised, the test environment could be used as a platform from which to attack production servers. Finally, the process of scanning for exposures is more of a spiral process than a linear process.

S3-41   Which of the following **BEST** ensures that modifications made to in-house developed business applications do not introduce new security exposures?

    A. Stress testing
    B. Patch management
    C. Change management
    D. Security baselines

C   Change management controls the process of introducing changes to systems to ensure that unintended changes are not introduced. Patch management involves the correction of software weaknesses and helps ensure that newly identified exploits are mitigated in a timely fashion. Security baselines provide minimum recommended settings. Stress testing ensures that there are no scalability problems.

S3-42   The advantage of Virtual Private Network (VPN) tunneling for remote users is that it:

    A. helps ensure that communications are secure.
    B. increases security between multi-tier systems.
    C. allows passwords to be changed less frequently.
    D. eliminates the need for secondary authentication.

A   Virtual Private Network (VPN) tunneling for remote users provides an encrypted link that helps ensure secure communications. It does not affect password change frequency, nor does it eliminate the need for secondary authentication or affect security within the internal network.

S3-43   Which of the following is **MOST** effective for securing wireless networks as a point of entry into a corporate network?

    A. Boundary router
    B. Strong encryption
    C. Internet-facing firewall
    D. Intrusion detection system (IDS)

B   Strong encryption is the most effective means of protecting wireless networks. Boundary routers, intrusion detection systems (IDSs) and firewalling the Internet would not be as effective.

S3-44   Which of the following is **MOST** effective in protecting against the attack technique known as phishing?

    A. Firewall blocking rules
    B. Up-to-date signature files
    C. Security awareness training
    D. Intrusion detection monitoring

C   Phishing relies on social engineering techniques. Providing good security awareness training will best reduce the likelihood of such an attack being successful. Firewall rules, signature files and intrusion detection system (IDS) monitoring will be largely unsuccessful at blocking this kind of attack.

S3-45   When a newly installed system for synchronizing passwords across multiple systems and platforms abnormally terminates without warning, which of the following should automatically occur **FIRST**?

    A. The firewall should block all inbound traffic during the outage
    B. All systems should block new logins until the problem is corrected
    C. Access control should fall back to nonsynchronized mode
    D. System logs should record all user activity for later analysis

C   The best mechanism is for the system to fallback to the original process of logging on individually to each system. Blocking traffic and new logins would be overly restrictive to the conduct of business, while recording all user activity would add little value.

S3-46   Which of the following is the **MOST** important risk associated with middleware in a client-server environment?

    A. Server patching may be prevented
    B. System backups may be incomplete
    C. System integrity may be affected
    D. End-user sessions may be hijacked

C   The major risk associated with middleware in a client-server environment is that system integrity may be adversely affected because of the very purpose of middleware, which is intended to support multiple operating environments interacting concurrently. Lack of proper software to control portability of data or programs across multiple platforms could result in a loss of data or program integrity. All other choices are less likely to occur.

S3-47   An outsource service provider must handle sensitive customer information. Which of the following is **MOST** important for an information security manager to know?

    A. Security in storage and transmission of sensitive data
    B. Provider's level of compliance with industry standards
    C. Security technologies in place at the facility
    D. Results of the latest independent security review

A   How the outsourcer protects the storage and transmission of sensitive information will allow an information security manager to understand how sensitive data will be protected. Choice B is an important but secondary consideration. Choice C is incorrect because security technologies are not the only components to protect the sensitive customer information. Choice D is incorrect because an independent security review may not include analysis on how sensitive customer information would be protected.

S3-48   Which of the following security mechanisms is **MOST** effective in protecting classified data that have been encrypted to prevent disclosure and transmission outside the organization's network?

    A. Configuration of firewalls
    B. Strength of encryption algorithms
    C. Authentication within application
    D. Safeguards over keys

D   If keys are in the wrong hands, documents will be able to be read regardless of where they are on the network. Choice A is incorrect because firewalls can be perfectly configured, but if the keys make it to the other side, they will not prevent the document from being decrypted. Choice B is incorrect because even easy encryption algorithms require adequate resources to break, whereas encryption keys can be easily used. Choice C is incorrect because the application "front door" controls may be bypassed by accessing data directly.

S3-49   In the process of deploying a new e-mail system, an information security manager would like to ensure the confidentiality of messages while in transit. Which of the following is the **MOST** appropriate method to ensure data confidentiality in a new e-mail system implementation?

   A.   Encryption
   B.   Digital certificate
   C.   Digital signature
   D.   Hashing algorithm

A   To preserve confidentiality of a message while in transit, encryption should be implemented. Choices B and C only help authenticate the sender and the receiver. Choice D ensures integrity.

S3-50   The **MOST** important reason that statistical anomaly-based intrusion detection systems (stat IDSs) are less commonly used than signature-based IDSs, is that stat IDSs:

   A.   create more overhead than signature-based IDSs.
   B.   cause false positives from minor changes to system variables.
   C.   generate false alarms from varying user or system actions.
   D.   cannot detect new types of attacks.

C   A statistical anomaly-based intrusion detection system (stat IDS) collects data from normal traffic and establishes a baseline. It then periodically samples the network activity based on statistical methods and compares samples to the baseline. When the activity is outside the baseline parameter (clipping level), the IDS notifies the administrator. The baseline variables can include a host's memory or central processing unit (CPU) usage, network packet types and packet quantities. If actions of the users or the systems on the network vary widely with periods of low activity and periods of frantic packet exchange, a stat IDS may not be suitable, as the dramatic swing from one level to another almost certainly will generate false alarms. This weakness will have the largest impact on the operation of the IT systems. Due to the nature of stat IDS operations (i.e., they must constantly attempt to match patterns of activity to the baseline parameters), a stat IDS requires much more overhead and processing than signature-based versions. Due to the nature of a stat IDS—based on statistics and comparing data with baseline parameters—this type of IDS may not detect minor changes to system variables and may generate many false positives. Choice D is incorrect; since the stat IDS can monitor multiple system variables, it can detect new types of variables by tracing for abnormal activity of any kind.

S3-51   The **MOST** important success factor to design an effective IT security awareness program is to:

   A.   customize the content to the target audience.
   B.   ensure senior management is represented.
   C.   ensure that all the staff is trained.
   D.   avoid technical content but give concrete examples.

A   Awareness training can only be effective if it is customized to the expectations and needs of attendees. Needs will be quite different depending on the target audience and will vary between business managers, end users and IT staff; program content and the level of detail communicated will therefore be different. Other criteria are also important; however, the customization of content is the most important factor.

S3-52   Which of the following practices completely prevents a man-in-the-middle (MitM) attack between
             two hosts?

             A.   Use security tokens for authentication
             B.   Connect through an IPSec VPN
             C.   Use https with a server-side certificate
             D.   Enforce static media access control (MAC) addresses

B   IPSec effectively prevents man-in-the-middle (MitM) attacks by including source and destination IPs
             within the encrypted portion of the packet. The protocol is resilient to MitM attacks. Using token-based
             authentication does not prevent a MitM attack; however, it may help eliminate reusability of stolen
             cleartext credentials. An https session can be intercepted through Domain Name Server (DNS) or Address
             Resolution Protocol (ARP) poisoning. ARP poisoning—a specific kind of MitM attack—may be prevented
             by setting static media access control (MAC) addresses. Nevertheless, DNS and NetBIOS resolution can
             still be attacked to deviate traffic.

S3-53   Which of the following features is normally missing when using Secure Sockets Layer (SSL) in a
             web browser?

             A.   Certificate-based authentication of web client
             B.   Certificate-based authentication of web server
             C.   Data confidentiality between client and web server
             D.   Multiple encryption algorithms

A   Web browsers have the capability of authenticating through client-based certificates; nevertheless, it is not
             commonly used. When using https, servers always authenticate with a certificate and, once the connection
             is established, confidentiality will be maintained between client and server. By default, web browsers and
             servers support multiple encryption algorithms and negotiate the best option upon connection.

S3-54   The **BEST** protocol to ensure confidentiality of transmissions in a business-to-customer (B2C) financial
             web application is:

             A.   Secure Sockets Layer (SSL).
             B.   Secure Shell (SSH).
             C.   IP Security (IPSec).
             D.   Secure/Multipurpose Internet Mail Extensions (S/MIME).

A   Secure Sockets Layer (SSL) is a cryptographic protocol that provides secure communications providing end
             point authentication and communications privacy over the Internet. In typical use, all data transmitted between
             the customer and the business are, therefore, encrypted by the business's web server and remain confidential.
             SSH File Transfer Protocol (SFTP) is a network protocol that provides file transfer and manipulation
             functionality over any reliable data stream. It is typically used with the SSH-2 protocol to provide secure file
             transfer. IP Security (IPSec) is a standardized framework for securing Internet Protocol (IP) communications
             by encrypting and/or authenticating each IP packet in a data stream. There are two modes of IPSec operation:
             transport mode and tunnel mode. Secure/Multipurpose Internet Mail Extensions (S/MIME) is a standard for
             public key encryption and signing of e-mail encapsulated in MIME; it is not a web transaction protocol.

S3-55   A message that has been encrypted by the sender's private key and again by the receiver's public key achieves:

    A. authentication and authorization.
    B. confidentiality and integrity.
    C. confidentiality and nonrepudiation.
    D. authentication and nonrepudiation.

C   Encryption by the private key of the sender will guarantee authentication and nonrepudiation. Encryption by the public key of the receiver will guarantee confidentiality.

S3-56   When a user employs a client-side digital certificate to authenticate to a web server through Secure Socket Layer (SSL), confidentiality is **MOST** vulnerable to which of the following?

    A. IP spoofing
    B. Man-in-the-middle attack
    C. Repudiation
    D. Trojan

D   A Trojan is a program that gives the attacker full control over the infected computer, thus allowing the attacker to hijack, copy or alter information after authentication by the user. IP spoofing will not work because IP is not used as an authentication mechanism. Man-in-the-middle attacks are not possible if using SSL with client-side certificates. Repudiation is unlikely because client-side certificates authenticate the user.

S3-57   Which of the following is the **MOST** relevant metric to include in an information security quarterly report to the executive committee?

    A. Security compliant servers trend report
    B. Percentage of security compliant servers
    C. Number of security patches applied
    D. Security patches applied trend report

A   The percentage of compliant servers will be a relevant indicator of the risk exposure of the infrastructure. However, the percentage is less relevant than the overall trend, which would provide a measurement of the efficiency of the IT security program. The number of patches applied would be less relevant, as this would depend on the number of vulnerabilities identified and patches provided by vendors.

S3-58   It is important to develop an information security baseline because it helps to define:

    A. critical information resources needing protection.
    B. a security policy for the entire organization.
    C. the minimum acceptable security to be implemented.
    D. required physical and logical access controls.

C   Developing an information security baseline helps to define the minimum acceptable security that will be implemented to protect the information resources in accordance with the respective criticality levels. Before determining the security baseline, an information security manager must establish the security policy, identify criticality levels of organization's information resources and assess the risk environment in which those resources operate.

S3-59   Which of the following **BEST** provides message integrity, sender identity authentication and nonrepudiation?

A.   Symmetric cryptography
B.   Public key infrastructure (PKI)
C.   Message hashing
D.   Message authentication code

B   Public key infrastructure (PKI) combines public key encryption with a trusted third party to publish and revoke digital certificates that contain the public key of the sender. Senders can digitally sign a message with their private key and attach their digital certificate (provided by the trusted third party). These characteristics allow senders to provide authentication, integrity validation and nonrepudiation. Symmetric cryptography provides confidentiality. Hashing can provide integrity and confidentiality. Message authentication codes provide integrity.

S3-60   Which of the following controls is **MOST** effective in providing reasonable assurance of physical access compliance to an unmanned server room controlled with biometric devices?

A.   Regular review of access control lists
B.   Security guard escort of visitors
C.   Visitor registry log at the door
D.   A biometric coupled with a PIN

A   A review of access control lists is a detective control that will enable an information security manager to ensure that authorized persons are entering in compliance with corporate policy. Visitors accompanied by a guard will also provide assurance but may not be cost effective. A visitor registry is the next cost-effective control. A biometric coupled with a PIN will strengthen the access control; however, compliance assurance logs will still have to be reviewed.

S3-61   To **BEST** improve the alignment of the information security objectives in an organization, the chief information security officer (CISO) should:

A.   revise the information security program.
B.   evaluate a balanced business scorecard.
C.   conduct regular user awareness sessions.
D.   perform penetration tests.

B   The balanced business scorecard can track the effectiveness of how an organization executes it information security strategy and determine areas of improvement. Revising the information security program may be a solution, but is not the best solution to improve alignment of the information security objectives. User awareness is just one of the areas the organization must track through the balanced business scorecard. Performing penetration tests does not affect alignment with information security objectives.

S3-62   When considering the value of assets, which of the following would give the information security manager the **MOST** objective basis for measurement of value delivery in information security governance?

A.   Number of controls
B.   Cost of achieving control objectives
C.   Effectiveness of controls
D.   Test results of controls

B   Comparison of cost of achievement of control objectives and corresponding value of assets sought to be protected would provide a sound basis for the information security manager to measure value delivery. Number of controls has no correlation with the value of assets unless the effectiveness of the controls and their cost are also evaluated. Effectiveness of controls has no correlation with the value of assets unless their costs are also evaluated. Test results of controls have no correlation with the value of assets unless the effectiveness of the controls and their cost are also evaluated.

S3-63   Which of the following, using public key cryptography, ensures authentication, confidentiality and nonrepudiation of a message?

   A.   Encrypting first by receiver's private key and second by sender's public key
   B.   Encrypting first by sender's private key and second by receiver's public key
   C.   Encrypting first by sender's private key and second decrypting by sender's public key
   D.   Encrypting first by sender's public key and second by receiver's private key

B   Encrypting by the sender's private key ensures authentication. By being able to decrypt with the sender's public key, the receiver would know that the message is sent by the sender only and the sender cannot deny/repudiate the message. By encrypting with the sender's public key secondly, only the sender will be able to decrypt the message and confidentiality is assured. The receiver's private key is private to the receiver and the sender cannot have it for encryption. Similarly, the receiver will not have the private key of the sender to decrypt the second-level encryption. In the case of encrypting first by the sender's private key and, second, decrypting by the sender's public key, confidentiality is not ensured since the message can be decrypted by anyone using the sender's public key. The receiver's private key would not be available to the sender for second-level encryption. Similarly, the sender's private key would not be available to the receiver for decrypting the message.

S3-64   A test plan to validate the security controls of a new system should be developed during which phase of the project?

   A.   Testing
   B.   Initiation
   C.   Design
   D.   Development

C   In the design phase, security checkpoints are defined and a test plan is developed. The testing phase is too late since the system has already been developed and is in production testing. In the initiation phase, the basic security objective of the project is acknowledged. Development is the coding phase and is too late to consider test plans.

S3-65   The **MOST** effective way to ensure that outsourced service providers comply with the organization's information security policy would be:

   A.   service level monitoring.
   B.   penetration testing.
   C.   periodically auditing.
   D.   security awareness training.

C   Regular audit exercise can spot any gap in the information security compliance. Service level monitoring can only pinpoint operational issues in the organization's operational environment. Penetration testing can identify security vulnerability but cannot ensure information compliance. Training can increase users' awareness on the information security policy, but is not more effective than auditing.

S3-66   In order to protect a network against unauthorized external connections to corporate systems, the information security manager should **BEST** implement:

    A.   a strong authentication.
    B.   IP antispoofing filtering.
    C.   network encryption protocol.
    D.   access lists of trusted devices.

A   Strong authentication will provide adequate assurance on the identity of the users, while IP antispoofing is aimed at the device rather than the user. Encryption protocol ensures data confidentiality and authenticity while access lists of trusted devices are easily exploited by spoofed identity of the clients.

S3-67   The **PRIMARY** driver to obtain external resources to execute the information security program is that external resources can:

    A.   contribute cost-effective expertise not available internally.
    B.   be made responsible for meeting the security program requirements.
    C.   replace the dependence on internal resources.
    D.   deliver more effectively on account of their knowledge.

A   Choice A represents the primary driver for the information security manager to make use of external resources. The information security manager will continue to be responsible for meeting the security program requirements despite using the services of external resources. The external resources should never completely replace the role of internal resources from a strategic perspective. The external resources cannot have a better knowledge of the business of the information security manager's organization than do the internal resources.

S3-68   The **MAIN** reason for deploying a public key infrastructure (PKI) when implementing an information security program is to:

    A.   ensure the confidentiality of sensitive material.
    B.   provide a high assurance of identity.
    C.   allow deployment of the active directory.
    D.   implement secure sockets layer (SSL) encryption.

B   The primary purpose of a public key infrastructure (PKI) is to provide strong authentication. Confidentiality is a function of the session keys distributed by the PKI. An active directory can use PKI for authentication as well as using other means. Even though secure sockets layer (SSL) encryption requires keys to authenticate, it is not the main reason for deploying PKI.

S3-69   Which of the following controls would **BEST** prevent accidental system shutdown from the console or operations area?

    A.   Redundant power supplies
    B.   Protective switch covers
    C.   Shutdown alarms
    D.   Biometric readers

B   Protective switch covers would reduce the possibility of an individual accidentally pressing the power button on a device, thereby turning off the device. Redundant power supplies would not prevent an individual from powering down a device. Shutdown alarms would be after the fact. Biometric readers would be used to control access to the systems.

S3-70   Which of the following is the **MOST** important reason that information security objectives should be defined?

    A. Tool for measuring effectiveness
    B. General understanding of goals
    C. Consistency with applicable standards
    D. Management sign-off and support initiatives

A   The creation of objectives can be used in part as a source of measurement of the effectiveness of information security management, which feeds into the overall governance. General understanding of goals and consistency with applicable standards are useful, but are not the primary reasons for having clearly defined objectives. Gaining management understanding is important, but by itself will not provide the structure for governance.

S3-71   What is the **BEST** policy for securing data on mobile universal serial bus (USB) drives?

    A. Authentication
    B. Encryption
    C. Prohibit employees from copying data to USB devices
    D. Limit the use of USB devices

B   Encryption provides the most effective protection of data on mobile devices. Authentication on its own is not very secure. Prohibiting employees from copying data to USB devices and limiting the use of USB devices are after the fact.

S3-72   When speaking to an organization's human resources department about information security, an information security manager should focus on the need for:

    A. an adequate budget for the security program.
    B. recruitment of technical IT employees.
    C. periodic risk assessments.
    D. security awareness training for employees.

D   An information security manager has to impress upon the human resources department the need for security awareness training for all employees. Budget considerations are more of an accounting function. The human resources department would become involved once they are convinced for the need of security awareness training. Recruiting IT-savvy staff may bring in new employees with better awareness of information security, but that is not a replacement for the training requirements of the other employees. Periodic risk assessments may or may not involve the human resources department function.

S3-73   Which of the following would **BEST** protect an organization's confidential data stored on a laptop computer from unauthorized access?

    A. Strong authentication by password
    B. Encrypted hard drives
    C. Multifactor authentication procedures
    D. Network-based data backup

B   Encryption of the hard disks will prevent unauthorized access to the laptop even when the laptop is lost or stolen. Strong authentication by password can be bypassed by a determined hacker. Multifactor authentication can be bypassed by removal of the hard drive and insertion into another laptop. Network-based data backups do not prevent access but rather recovery from data loss.

S3-74   What is the **MOST** important reason for conducting security awareness programs throughout an organization?

A. Reducing the human risk
B. Maintaining evidence of training records to ensure compliance
C. Informing business units about the security strategy
D. Training personnel in security incident response

A   People are the weakest link in security implementation, and awareness would reduce this risk. Through security awareness and training programs, individual employees can be informed and sensitized on various security policies and other security topics, thus ensuring compliance from each individual. Laws and regulations also aim to reduce human risk. Informing business units about the security strategy is best done through steering committee meetings or other forums.

S3-75   At what stage of the applications development process would encryption key management initially be addressed?

A. Requirements development
B. Deployment
C. Systems testing
D. Code reviews

A   Encryption key management has to be integrated into the requirements of the application's design. During systems testing and deployment would be too late since the requirements have already been agreed upon. Code reviews are part of the final quality assurance (QA) process and would also be too late in the process.

S3-76   The **MOST** effective way to ensure network users are aware of their responsibilities to comply with an organization's security requirements is:

A. messages displayed at every logon.
B. periodic security-related e-mail messages.
C. an Intranet web site for information security.
D. circulating the information security policy.

A   Logon banners would appear every time the user logs on, and the user would be required to read and agree to the same before using the resources. Also, as the message is conveyed in writing and appears consistently, it can be easily enforceable in any organization. Security-related e-mail messages are frequently considered as "Spam" by network users and do not, by themselves, ensure that the user agrees to comply with security requirements. The existence of an Intranet web site does not force users to access it and read the information. Circulating the information security policy alone does not confirm that an individual user has read, understood and agreed to comply with its requirements unless it is associated with formal acknowledgment, such as a user's signature of acceptance.

S3-77   Which of the following would be the **BEST** defense against sniffing?

A. Password protect the files
B. Implement a dynamic IP address scheme
C. Encrypt the data being transmitted
D. Set static mandatory access control (MAC) addresses

C   Encrypting the data will obfuscate the data so that they are not visible in plain text. Someone would have to collate the entire data stream and try decrypting it, which is not easy. Passwords can be recovered by brute-force attacks and by password crackers, so this is not the best defense against sniffing. IP addresses can always be discovered, even if dynamic IP addresses are implemented. The person sniffing traffic can initiate multiple sessions for possible IP addresses. Setting static mandatory access control (MAC) addresses can prevent address resolution protocol (ARP) poisoning, but it does not prevent sniffing.

S3-78   A digital signature using a public key infrastructure (PKI) will:

    A. not ensure the integrity of a message.
    B. rely on the extent to which the certificate authority (CA) is trusted.
    C. require two parties to the message exchange.
    D. provide a high level of confidentiality.

B   The certificate authority (CA) is a trusted third party that attests to the identity of the signatory, and reliance will be a function of the level of trust afforded the CA. A digital signature would provide a level of assurance of message integrity, but it is a three-party exchange, including the CA. Digital signatures do not require encryption of the message in order to preserve confidentiality.

S3-79   When configuring a biometric access control system that protects a high-security data center, the system's sensitivity level should be set:

    A. to a higher false reject rate (FRR).
    B. to a lower crossover error rate.
    C. to a higher false acceptance rate (FAR).
    D. exactly to the crossover error rate.

A   Biometric access control systems are not infallible. When tuning the solution, one has to adjust the sensitivity level to give preference either to false reject rate (type I error rate) where the system will be more prone to err denying access to a valid user or erring and allowing access to an invalid user. As the sensitivity of the biometric system is adjusted, these values change inversely. At one point, the two values intersect and are equal. This condition creates the crossover error rate, which is a measure of the system accuracy. In systems where the possibility of false rejects is a problem, it may be necessary to reduce sensitivity and thereby increase the number of false accepts. This is sometimes referred to as equal error rate (EER). In a very sensitive system, it may be desirable to minimize the number of false accepts—the number of unauthorized persons allowed access. To do this, the system is tuned to be more sensitive, which causes the false rejects—the number of authorized persons disallowed access—to increase.

S3-80   An organization has adopted a practice of regular staff rotation to minimize the risk of fraud and encourage crosstraining. Which type of authorization policy would **BEST** address this practice?

    A. Multilevel
    B. Role-based
    C. Discretionary
    D. Attribute-based

B   A role-based policy will associate data access with the role performed by an individual, thus restricting access to data required to perform the individual's tasks. Multilevel policies are based on classifications and clearances. Discretionary policies leave access decisions up to information resource managers.

S3-81   Which of the following is the **MOST** important reason for an information security review of contracts? To help ensure that:

A. the parties to the agreement can perform.
B. confidential data are not included in the agreement.
C. appropriate controls are included.
D. the right to audit is a requirement.

C   Agreements with external parties can expose an organization to information security risks that must be assessed and appropriately mitigated. The ability of the parties to perform is normally the responsibility of legal and the business operation involved. Confidential information may be in the agreement by necessity and, while the information security manager can advise and provide approaches to protect the information, the responsibility rests with the business and legal. Audit rights may be one of many possible controls to include in a third-party agreement, but is not necessarily a contract requirement, depending on the nature of the agreement.

S3-82   For virtual private network (VPN) access to the corporate network, the information security manager is requiring strong authentication. Which of the following is the strongest method to ensure that logging onto the network is secure?

A. Biometrics
B. Symmetric encryption keys
C. Secure Sockets Layer (SSL)-based authentication
D. Two-factor authentication

D   Two-factor authentication requires more than one type of user authentication. While biometrics provides unique authentication, it is not strong by itself, unless a PIN or some other authentication factor is used with it. Biometric authentication by itself is also subject to replay attacks. A symmetric encryption method that uses the same secret key to encrypt and decrypt data is not a typical authentication mechanism for end users. This private key could still be compromised. SSL is the standard security technology for establishing an encrypted link between a web server and a browser. SSL is not an authentication mechanism. If SSL is used with a client certificate and a password, it would be a two-factor authentication.

S3-83   An organization's information security manager is planning the structure of the Information Security Steering Committee. Which of the following groups should the manager invite?

A. External audit and network penetration testers
B. Board of directors and the organization's regulators
C. External trade union representatives and key security vendors
D. Leadership from IT, human resources and the sales department

D   Leaders from IT, human resources and sales are key individuals who must support an information security program. External audit may assess and advise on the program, and testers may be used by the program, but they are not appropriate steering committee members. The steering committee needs to have practitioner-level representation. It may report to the board, but board members would not generally be part of the steering committee, except for its executive sponsor. Regulators would not participate on this committee. External trade union representatives and key security vendors are entities that may need to be consulted as part of program activities, but would not be members of the steering committee.

S3-84   Which of the following is the **MOST** effective security measure to protect data held on mobile computing devices?

    A. Biometric access control
    B. Encryption of stored data
    C. Power-on passwords
    D. Protection of data being transmitted

B   Encryption of stored data will help ensure that the actual data cannot be recovered without the encryption key. Biometric access control does not necessarily protect stored data. Choice C is incorrect because power-on passwords do not protect data effectively. Choice D is incorrect because it relates to protecting data in transmission.

S3-85   Which of the following is **MOST** useful in managing increasingly complex security deployments?

    A. A standards-based approach
    B. A security architecture
    C. Policy development
    D. Senior management support

B   Deploying complex security initiatives and integrating a range of diverse projects and activities would be more easily managed with the overview and relationships provided by a security architecture. Standards may provide metrics for deployment and policies would guide direction, but standards and policies would not provide significant management tools.

S3-86   An organization is implementing intrusion protection in their demilitarized zone (DMZ). Which of the following steps is necessary to make sure that the intrusion prevention system (IPS) can view all traffic in the DMZ? Ensure that:

    A. intrusion prevention is placed in front of the firewall.
    B. all devices that are connected can easily see the IPS in the network.
    C. all encrypted traffic is decrypted prior to being processed by the IPS.
    D. traffic to all devices is mirrored to the IPS.

C   All encryption should be terminated to allow all traffic to be viewed by the IPS. The encryption should be terminated at a hardware Secure Sockets Layer (SSL) accelerator or virtual private network (VPN) server to allow all traffic to be monitored since encrypted traffic is unreadable.

S3-87   Which of the following guarantees that data in a file have not changed?

    A. Inspecting the modified date of the file
    B. Encrypting the file with symmetric encryption
    C. Using stringent access control to prevent unauthorized access
    D. Creating a hash of the file, then comparing the file hashes

D   A hashing algorithm can be used to mathematically ensure that data haven't been changed by hashing a file and comparing the hashes after a suspected change.

S3-88   Which of the following mechanisms is the **MOST** secure way to implement a secure wireless network?

     A. Filter media access control (MAC) addresses
     B. Use a Wi-Fi Protected Access (WPA2) protocol
     C. Use a Wired Equivalent Privacy (WEP) key
     D. Web-based authentication

B   WPA2 is currently one of the most secure authentication and encryption protocols for mainstream wireless products. MAC address filtering by itself is not a good security mechanism since allowed MAC addresses can be easily sniffed and then spoofed to get into the network. WEP is no longer a secure encryption mechanism for wireless communications. The WEP key can be easily broken within minutes using widely available software. And once the WEP key is obtained, all communications of every other wireless client are exposed. Finally, a web-based authentication mechanism can be used to prevent unauthorized user access to a network, but it will not solve the wireless network's main security issues, such as preventing network sniffing.

S3-89   Which of the following devices could potentially stop a Structured Query Language (SQL) injection attack?

     A. An intrusion prevention system (IPS)
     B. An intrusion detection system (IDS)
     C. A host-based intrusion detection system (HIDS)
     D. A host-based firewall

A   SQL injection attacks occur at the application layer. Most IPS vendors will detect at least basic sets of SQL injection and will be able to stop them. IDS will detect, but not prevent. HIDS will be unaware of SQL injection problems. A host-based firewall, be it on the web server or the database server, will allow the connection because firewalls do not check packets at an application layer.

S3-90   Nonrepudiation can **BEST** be ensured by using:

     A. strong passwords.
     B. a digital hash.
     C. symmetric encryption.
     D. digital signatures.

D   Digital signatures use a private and public key pair, authenticating both parties. The integrity of the contents exchanged is controlled through the hashing mechanism that is signed by the private key of the exchanging party. A digital hash in itself helps in ensuring integrity of the contents, but not nonrepudiation. Symmetric encryption wouldn't help in nonrepudiation since the keys are always shared between parties. Strong passwords only ensure authentication to the system and cannot be used for nonrepudiation involving two or more parties.

S3-91   Which of the following is the **MOST** important step before implementing a security policy?

     A. Communicating to employees
     B. Training IT staff
     C. Identifying relevant technologies for automation
     D. Obtaining sign-off from stakeholders

D   Sign-off must be obtained from all stakeholders since that would signify formal acceptance of all the policy objectives and expectations of the business along with all residual risks. Only after sign-off is obtained can the other mentioned activities begin.

S3-92   An organization's security awareness program should focus on which of the following?

    A. Establishing metrics for network backups
    B. Installing training software which simulates security incidents
    C. Communicating what employees should do/not do in the context of their job responsibilities
    D. Accessing levels within the organization for applications and the Internet

C   An organization's security awareness program should focus on employee behavior and the consequences of both compliance and noncompliance with the security policy.

S3-93   The **BEST** defense against successful phishing attacks is:

    A. application hardening.
    B. spam filters.
    C. an intrusion detection system (IDS).
    D. end user awareness.

D   Phishing attacks are due to social engineering attacks and are best defended by user awareness training. Application hardening, spam filters and IDSs are inadequate since the phishing attacks usually don't have the same patterns or unique signatures.

S3-94   Which of the following would be the **GREATEST** challenge when developing a standard awareness training program for a global organization?

    A. Technical input requirements for IT security staff
    B. Evaluating training program effectiveness
    C. A diverse culture and varied technical abilities of end users
    D. Availability of users either on weekends or after office hours

C   A diverse culture and differences in the levels of IT knowledge and IT exposure pose the most difficulties when developing a standard training program since the learning needs of employees vary. IT security staff will require technical inputs and having a separate training program would not be considered a challenge. Evaluating training program effectiveness is not a problem when developing a standard training program. In fact, the evaluation of training program effectiveness will be easier for a standard training program delivered across the organization. Availability of users on weekends or beyond office hours has no impact on the development of a standard training program.

S3-95   When outsourcing, in order to ensure that third-party service providers comply with an organization security policy, which of the following should occur?

    A. A predefined meeting schedule
    B. A periodic security audit
    C. Inclusion in the contract of a list of individuals to be called in the event of an incident (call tree)
    D. Inclusion in the contract of a confidentiality clause

B   A periodic security audit is a formal and documented way to promote compliance. All other options are contributors to, but do not ensure, compliance.

S3-96   Which of the following security controls addresses availability?

    A. Least privilege
    B. Public key infrastructure
    C. Role-based access
    D. Contingency planning

D   Contingency planning ensures that the system and data are available in the event of a problem. Choices A, B and C are not correct because least privilege is an access control that is concerned with confidentiality, public key infrastructure with confidentiality and integrity, and role-based access with integrity and confidentiality, not integrity alone.

S3-97   Which of the following control measures **BEST** addresses integrity?

    A. Nonrepudiation
    B. Timestamps
    C. Biometric scanning
    D. Encryption

A   Nonrepudiation is a control technique that addresses the integrity of information by ensuring that the originator of a message or transaction cannot repudiate (deny or reject) the message, so the message or transaction can be considered authorized, authentic and valid. The other choices are not correct. Using timestamps is a control that addresses only one component of message integrity. Biometric scanning is a control that addresses confidentiality. Encryption is a control that addresses confidentiality, and may be an element of a data integrity scheme, but is not sufficient to achieve the same level of integrity as the set of measures used to ensure nonrepudiation.

S3-98   The **BEST** way to ensure that security settings on each platform are in compliance with information security policies and procedures is to:

    A. perform penetration testing.
    B. establish security baselines.
    C. implement vendor default settings.
    D. link policies to an independent standard.

B   Security baselines will provide the best assurance that each platform meets minimum criteria. Penetration testing will not be as effective and can only be performed periodically. Vendor default settings will not necessarily meet the criteria set by the security policies, while linking policies to an independent standard will not provide assurance that the platforms meet these levels of security.

S3-99   A web-based business application is being migrated from test to production. Which of the following is the **MOST** important management signoff for this migration?

    A. User
    B. Network
    C. Operations
    D. Database

A   As owners of the system, user management signoff is the most important. If a system does not meet the needs of the business, then it has not met its primary objective. The needs of network, operations and database management are secondary to the needs of the business.

CISM Review Questions, Answers & Explanations Manual 2012
ISACA. All Rights Reserved.

S3-100   The **BEST** way to ensure that information security policies are followed is to:

    A. distribute printed copies to all employees.
    B. perform periodic reviews for compliance.
    C. include escalating penalties for noncompliance.
    D. establish an anonymous hotline to report policy abuses.

B   The best way to ensure that information security policies are followed is to periodically review levels of compliance. Distributing printed copies, advertising an abuse hotline or linking policies to an international standard will not motivate individuals as much as the consequences of being found in noncompliance. Escalating penalties will first require a compliance review.

S3-101   The **MOST** appropriate individual to determine the level of information security needed for a specific business application is the:

    A. system developer.
    B. information security manager.
    C. steering committee.
    D. system data owner.

D   Data owners are the most knowledgeable of the security needs of the business application for which they are responsible. The system developer, security manager and system custodian will have specific knowledge on limited areas but will not have full knowledge of the business issues that affect the level of security required. The steering committee does not perform at that level of detail on the operation.

S3-102   Which of the following will **MOST** likely reduce the chances of an unauthorized individual gaining access to computing resources by pretending to be an authorized individual needing to have his/her password reset?

    A. Performing reviews of password resets
    B. Conducting security awareness programs
    C. Increasing the frequency of password changes
    D. Implementing automatic password syntax checking

B   Social engineering can be mitigated best through periodic security awareness training for staff members who may be the target of such an attempt. Changing the frequency of password changes, strengthening passwords and checking the number of password resets may be desirable, but they will not be as effective in reducing the likelihood of a social engineering attack.

S3-103   Which of the following is the **MOST** likely to change an organization's culture to one that is more security conscious?

    A. Adequate security policies and procedures
    B. Periodic compliance reviews
    C. Security steering committees
    D. Security awareness campaigns

D   Security awareness campaigns will be more effective at changing an organizational culture than the creation of steering committees and security policies and procedures. Compliance reviews are helpful; however, awareness by all staff is more effective because compliance reviews are focused on certain areas/groups and do not necessarily educate.

S3-104   The **BEST** way to ensure that an external service provider complies with organizational security policies is to:

A. explicitly include the service provider in the security policies.
B. receive acknowledgment in writing stating the provider has read all policies.
C. cross-reference to policies in the service level agreement.
D. perform periodic reviews of the service provider.

D   Periodic reviews will be the most effective way of obtaining compliance from the external service provider. References in policies and service level agreements and requesting written acknowledgement will not be as effective since they will not trigger the detection of noncompliance.

S3-105   When an emergency security patch is received via electronic mail, the patch should **FIRST** be:

A. loaded onto an isolated test machine.
B. decompiled to check for malicious code.
C. validated to ensure its authenticity.
D. copied onto write-once media to prevent tampering.

C   It is important to first validate that the patch is authentic. Only then should it be copied onto write-once media, decompiled to check for malicious code or loaded onto an isolated test machine.

S3-106   In a well-controlled environment, which of the following activities is **MOST** likely to lead to the introduction of weaknesses in security software?

A. Applying patches
B. Changing access rules
C. Upgrading hardware
D. Backing up files

B   Security software will generally have a well-controlled process for applying patches, backing up files and upgrading hardware. The greatest risk occurs when access rules are changed since they are susceptible to being opened up too much, which can result in the creation of a security exposure.

S3-107   Which of the following is the **BEST** indicator that security awareness training has been effective?

A. Employees sign to acknowledge the security policy
B. More incidents are being reported
C. A majority of employees have completed training
D. No incidents have been reported in three months

B   More incidents being reported could be an indicator that the staff is paying more attention to security. Employee signatures and training completion may or may not have anything to do with awareness levels. The number of individuals trained may not indicate they are more aware. No recent security incidents does not reflect awareness levels, but may prompt further research to confirm.

S3-108   Which of the following metrics would be the **MOST** useful in measuring how well information security is monitoring violation logs?

    A. Penetration attempts investigated
    B. Violation log reports produced
    C. Violation log entries
    D. Frequency of corrective actions taken

A   The most useful metric is one that measures the degree to which complete follow-through has taken place. The quantity of reports, entries on reports and the frequency of corrective actions are not indicative of whether or not investigative action was taken.

S3-109   Which of the following change management activities would be a clear indicator that normal operational procedures require examination? A high percentage of:

    A. similar change requests.
    B. change request postponements.
    C. canceled change requests.
    D. emergency change requests.

D   A high percentage of emergency change requests could be caused by changes that are being introduced at the last minute to bypass normal change management procedures. Similar requests, postponements and canceled requests all are indicative of a properly functioning change management process.

S3-110   Which of the following is the **MOST** important management signoff for migrating an order processing system from a test environment to a production environment?

    A. User
    B. Security
    C. Operations
    D. Database

A   As owners of the system, user management approval would be the most important. Although the signoffs of security, operations and database management may be appropriate, they are secondary to ensuring the new system meets the requirements of the business.

S3-111   Prior to having a third party perform an attack and penetration test against an organization, the **MOST** important action is to ensure that:

    A. the third party provides a demonstration on a test system.
    B. goals and objectives are clearly defined.
    C. the technical staff has been briefed on what to expect.
    D. special backups of production servers are taken.

B   The most important action is to clearly define the goals and objectives of the test. Assuming that adequate backup procedures are in place, special backups should not be necessary. Technical staff should not be briefed nor should there be a demo as this will reduce the spontaneity of the test.

S3-112   When a departmental system continues to be out of compliance with an information security policy's password strength requirements, the **BEST** action to undertake is to:

A.   submit the issue to the steering committee.
B.   conduct a risk assessment to quantify the risk.
C.   isolate the system from the rest of the network.
D.   request a risk acceptance from senior management.

B   A risk assessment is warranted to determine whether a risk acceptance should be granted and to demonstrate to the department the danger of deviating from the established policy. Isolating the system would not support the needs of the business. Any waiver should be granted only after performing a risk assessment.

S3-113   Which of the following is **MOST** important to the successful promotion of good security management practices?

A.   Security metrics
B.   Security baselines
C.   Management support
D.   Periodic training

C   Without management support, all other efforts will be undermined. Metrics, baselines and training are all important, but they depend on management support for their success.

S3-114   Which of the following environments represents the **GREATEST** risk to organizational security?

A.   Locally managed file server
B.   Enterprise data warehouse
C.   Load-balanced, web server cluster
D.   Centrally managed data switch

A   A locally managed file server will be the least likely to conform to organizational security policies because it is generally subject to less oversight and monitoring. Centrally managed data switches, web server clusters and data warehouses are subject to close scrutiny, good change control practices and monitoring.

S3-115   Nonrepudiation can **BEST** be assured by using:

A.   delivery path tracing.
B.   reverse lookup translation.
C.   out-of-band channels.
D.   digital signatures.

D   Effective nonrepudiation requires the use of digital signatures. Reverse lookup translation involves converting Internet Protocol (IP) addresses to usernames. Delivery path tracing shows the route taken but does not confirm the identity of the sender. Out-of-band channels are useful when, for confidentiality, it is necessary to break a message into two parts that are sent by different means.

S3-116   Of the following, the **BEST** method for ensuring that temporary employees do not receive excessive access rights is:

    A. mandatory access controls.
    B. discretionary access controls.
    C. lattice-based access controls.
    D. role-based access controls.

D   Role-based access controls will grant temporary employee access based on the job function to be performed. This provides a better means of ensuring that the access is not more or less than what is required. Discretionary, mandatory and lattice-based access controls are all security models, but they do not address the issue of temporary employees as well as role-based access controls.

S3-117   Which of the following areas is **MOST** susceptible to the introduction of security weaknesses?

    A. Database management
    B. Tape backup management
    C. Configuration management
    D. Incident response management

C   Configuration management provides the greatest likelihood of security weaknesses through misconfiguration and failure to update operating system (OS) code correctly and on a timely basis.

S3-118   Security policies should be aligned **MOST** closely with:

    A. industry best practices.
    B. organizational needs.
    C. generally accepted standards.
    D. local laws and regulations.

B   The needs of the organization should always take precedence. Best practices and local regulations are important, but they do not take into account the total needs of an organization.

S3-119   The **BEST** way to determine if an anomaly-based intrusion detection system (IDS) is properly installed is to:

    A. simulate an attack and review IDS performance.
    B. use a honeypot to check for unusual activity.
    C. audit the configuration of the IDS.
    D. benchmark the IDS against a peer site.

A   Simulating an attack on the network demonstrates whether the intrusion detection system (IDS) is properly tuned. Reviewing the configuration may or may not reveal weaknesses since an anomaly-based system uses trends to identify potential attacks. A honeypot is not a good first step since it would need to have already been penetrated. Benchmarking against a peer site would generally not be practical or useful.

S3-120   The **BEST** time to perform a penetration test is after:

A. an attempted penetration has occurred.
B. an audit has reported weaknesses in security controls.
C. various infrastructure changes are made.
D. a high turnover in systems staff.

C   Changes in the systems infrastructure are most likely to inadvertently introduce new exposures. Conducting a test after an attempted penetration is not as productive since an organization should not wait until it is attacked to test its defenses. Any exposure identified by an audit should be corrected before it would be appropriate to test. A turnover in administrative staff does not warrant a penetration test, although it may warrant a review of password change practices and configuration management.

S3-121   Successful social engineering attacks can **BEST** be prevented through:

A. preemployment screening.
B. close monitoring of users' access patterns.
C. periodic awareness training.
D. efficient termination procedures.

C   Security awareness training is most effective in preventing the success of social engineering attacks by providing users with the awareness they need to resist such attacks. Screening of new employees, monitoring and rapid termination will not be effective against external attacks.

S3-122   Which of the following is the **BEST** way to ensure that an intruder who successfully penetrates a network will be detected before significant damage is inflicted?

A. Perform periodic penetration testing
B. Establish minimum security baselines
C. Implement vendor default settings
D. Install a honeypot on the network

D   Honeypots attract hackers away from sensitive systems and files. Since honeypots are closely monitored, the intrusion is more likely to be detected before significant damage is inflicted. Security baselines will only provide assurance that each platform meets minimum criteria. Penetration testing is not as effective and can only be performed sporadically. Vendor default settings are not effective.

S3-123   Which of the following presents the **GREATEST** threat to the security of an enterprise resource planning (ERP) system?

A. User ad hoc reporting is not logged
B. Network traffic is through a single switch
C. Operating system (OS) security patches have not been applied
D. Database security defaults to ERP settings

C   The fact that operating system (OS) security patches have not been applied is a serious weakness. Routing network traffic through a single switch is not unusual. Although the lack of logging for user ad hoc reporting is not necessarily good, it does not represent as serious a security weakness as the failure to install security patches. Database security defaulting to the ERP system's settings is not as significant.

S3-124   In a social engineering scenario, which of the following will **MOST** likely reduce the likelihood of an unauthorized individual gaining access to computing resources?

    A. Implementing on-screen masking of passwords
    B. Conducting periodic security awareness programs
    C. Increasing the frequency of password changes
    D. Requiring that passwords be kept strictly confidential

B   Social engineering can best be mitigated through periodic security awareness training for users who may be the target of such an attempt. Implementing on-screen masking of passwords and increasing the frequency of password changes are desirable, but these will not be effective in reducing the likelihood of a successful social engineering attack. Requiring that passwords be kept secret in security policies is a good control but is not as effective as periodic security awareness programs that will alert users of the dangers posed by social engineering.

S3-125   Which of the following will **BEST** ensure that management takes ownership of the decision making process for information security?

    A. Security policies and procedures
    B. Annual self-assessment by management
    C. Security steering committees
    D. Security awareness campaigns

C   Security steering committees provide a forum for management to express its opinion and take ownership in the decision making process. Security awareness campaigns, security policies and procedures, and self-assessment exercises are all good but do not exemplify the taking of ownership by management.

S3-126   Which of the following is the **MOST** appropriate individual to implement and maintain the level of information security needed for a specific business application?

    A. System analyst
    B. Quality control manager
    C. Process owner
    D. Information security manager

C   Process owners implement information protection controls as determined by the business' needs. Process owners have the most knowledge about security requirements for the business application for which they are responsible. The system analyst, quality control manager, and information security manager do not possess the necessary knowledge or authority to implement and maintain the appropriate level of business security.

S3-127   Which of the following activities is **MOST** likely to increase the difficulty of totally eradicating malicious code that is not immediately detected?

    A. Applying patches
    B. Changing access rules
    C. Upgrading hardware
    D. Backing up files

D   If malicious code is not immediately detected, it will most likely be backed up as a part of the normal tape backup process. When later discovered, the code may be eradicated from the device but still remain undetected on a backup tape. Any subsequent restores using that tape may reintroduce the malicious code. Applying patches, changing access rules and upgrading hardware does not significantly increase the level of difficulty.

S3-128   Security awareness training should be provided to new employees:

A. on an as-needed basis.
B. during system user training.
C. before they have access to data.
D. along with department staff.

C   Security awareness training should occur before access is granted to ensure the new employee understands that security is part of the system and business process. All other choices imply that security awareness training is delivered subsequent to the granting of system access, which may place security as a secondary step.

S3-129   What is the **BEST** method to verify that all security patches applied to servers were properly documented?

A. Trace change control requests to operating system (OS) patch logs
B. Trace OS patch logs to OS vendor's update documentation
C. Trace OS patch logs to change control requests
D. Review change control documentation for key servers

C   To ensure that all patches applied went through the change control process, it is necessary to use the operating system (OS) patch logs as a starting point and then check to see if change control documents are on file for each of these changes. Tracing from the documentation to the patch log will not indicate if some patches were applied without being documented. Similarly, reviewing change control documents for key servers or comparing patches applied to those recommended by the OS vendor's web site does not confirm that these security patches were properly approved and documented.

S3-130   A security awareness program should:

A. present top management's perspective.
B. address details on specific exploits.
C. address specific groups and roles.
D. promote security department procedures.

C   Different groups of employees have different levels of technical understanding and need awareness training that is customized to their needs; it should not be presented from a specific perspective. Specific details on technical exploits should be avoided since this may provide individuals with knowledge they might misuse or it may confuse the audience. This is also not the best forum in which to present security department procedures.

S3-131   The **PRIMARY** objective of security awareness is to:

A. ensure that security policies are understood.
B. influence employee behavior.
C. ensure legal and regulatory compliance.
D. notify of actions for noncompliance.

B   It is most important that security-conscious behavior be encouraged among employees through training that influences expected responses to security incidents. Ensuring that policies are read and understood, giving employees fair warning of potential disciplinary action, or meeting legal and regulatory requirements is important but secondary.

S3-132   Which of the following will **BEST** protect against malicious activity by a former employee?

    A. Preemployment screening
    B. Close monitoring of users
    C. Periodic awareness training
    D. Effective termination procedures

D   When an employee leaves an organization, the former employee may attempt to use their credentials to perform unauthorized or malicious activity. Accordingly, it is important to ensure timely revocation of all access at the time an individual is terminated. Security awareness training, preemployment screening and monitoring are all important, but are not as effective in preventing this type of situation.

S3-133   Which of the following represents a **PRIMARY** area of interest when conducting a penetration test?

    A. Data mining
    B. Network mapping
    C. Intrusion Detection System (IDS)
    D. Customer data

B   Network mapping is the process of determining the topology of the network one wishes to penetrate. This is one of the first steps toward determining points of attack in a network. Data mining is associated with *ad hoc* reporting and, together with customer data, they are potential targets after the network is penetrated. The intrusion detection mechanism in place is not an area of focus because one of the objectives is to determine how effectively it protects the network or how easy it is to circumvent.

S3-134   The return on investment of information security can **BEST** be evaluated through which of the following?

    A. Support of business objectives
    B. Security metrics
    C. Security deliverables
    D. Process improvement models

A   One way to determine the return on security investment is to illustrate how information security supports the achievement of business objectives. Security metrics measure improvement and effectiveness within the security practice but do not tie to business objectives. Similarly, listing deliverables and creating process improvement models does not necessarily tie into business objectives.

S3-135   To help ensure that contract personnel do not obtain unauthorized access to sensitive information, an information security manager should **PRIMARILY**:

    A. set their accounts to expire in six months or less.
    B. avoid granting system administration roles.
    C. ensure they successfully pass background checks.
    D. ensure their access is approved by the data owner.

B   Contract personnel should not be given job duties that provide them with power user or other administrative roles that they could then use to grant themselves access to sensitive files. Setting expiration dates, requiring background checks and having the data owner assign access are all positive elements, but these will not prevent contract personnel from obtaining access to sensitive information.

S3-136   Security audit reviews should **PRIMARILY**:

    A. ensure that controls operate as required.
    B. ensure that controls are cost-effective.
    C. focus on preventive controls.
    D. ensure controls are technologically current.

A   The primary objective of a security review or audit should be to provide assurance on the adequacy of security controls. Reviews should focus on all forms of control, not just on preventive control. Cost-effectiveness and technological currency are important but not as critical.

S3-137   Which of the following is the **MOST** appropriate method to protect a password that opens a confidential file?

    A. Delivery path tracing
    B. Reverse lookup translation
    C. Out-of-band channels
    D. Digital signatures

C   Out-of-band channels are useful when it is necessary, for confidentiality, to break a message into two parts that are then sent by different means. Digital signatures only provide nonrepudiation. Reverse lookup translation involves converting an Internet Protocol (IP) address to a username. Delivery path tracing shows the route taken but does not confirm the identity of the sender.

S3-138   What is the **MOST** effective access control method to prevent users from sharing files with unauthorized users?

    A. Mandatory
    B. Discretionary
    C. Walled garden
    D. Role-based

A   Mandatory access controls restrict access to files based on the security classification of the file. This prevents users from sharing files with unauthorized users. Role-based access controls grant access according to the role assigned to a user; they do not prohibit file sharing. Discretionary and lattice-based access controls are not as effective as mandatory access controls in preventing file sharing. A walled garden is an environment that controls a user's access to web content and services. In effect, the walled garden directs the user's navigation within particular areas, and does not necessarily prevent sharing of other material.

S3-139   Which of the following is an inherent weakness of signature-based intrusion detection systems?

    A. A higher number of false positives
    B. New attack methods will be missed
    C. Long duration probing will be missed
    D. Attack profiles can be easily spoofed

B   Signature-based intrusion detection systems do not detect new attack methods for which signatures have not yet been developed. False positives are not necessarily any higher, and spoofing is not relevant in this case. Long duration probing is more likely to fool anomaly-based systems (boiling frog technique).

CISM Review Questions, Answers & Explanations Manual 2012
ISACA. All Rights Reserved.

S3-140 Which of the following is the **BEST** way to ensure that a corporate network is adequately secured against external attack?

    A. Utilize an intrusion detection system.
    B. Establish minimum security baselines.
    C. Implement vendor recommended settings.
    D. Perform periodic penetration testing.

D Penetration testing is the best way to assure that perimeter security is adequate. An intrusion detection system (IDS) may detect an attempted attack, but it will not confirm whether the perimeter is secure. Minimum security baselines and applying vendor recommended settings are beneficial, but they will not provide the level of assurance that is provided by penetration testing.

S3-141 Which of the following presents the **GREATEST** exposure to internal attack on a network?

    A. User passwords are not automatically expired
    B. All network traffic goes through a single switch
    C. User passwords are encoded but not encrypted
    D. All users reside on a single internal subnet

C When passwords are sent over the internal network in an encoded format, they can easily be converted to clear text. All passwords should be encrypted to provide adequate security. Not automatically expiring user passwords does create an exposure, but not as great as having unencrypted passwords. Using a single switch or subnet does not present a significant exposure.

S3-142 Which of the following provides the linkage to ensure that procedures are correctly aligned with information security policy requirements?

    A. Standards
    B. Guidelines
    C. Security metrics
    D. IT governance

A Standards are the bridge between high-level policy statements and the "how to" detailed format of procedures. Security metrics and governance would not ensure correct alignment between policies and procedures. Similarly, guidelines are not linkage documents but rather provide suggested guidance on best practices.

S3-143 Which of the following are the **MOST** important individuals to include as members of an information security steering committee?

    A. Direct reports to the chief information officer
    B. IT management and key business process owners
    C. Cross-section of end users and IT professionals
    D. Internal audit and corporate legal departments

B Security steering committees provide a forum for management to express its opinion and take some ownership in the decision making process. It is imperative that business process owners be included in this process. None of the other choices includes input by business process owners.

S3-144   Data owners are normally responsible for which of the following?

A. Applying emergency changes to application data
B. Administering security over database records
C. Migrating application code changes to production
D. Determining the level of application security required

D   Data owners approve access to data and determine the degree of protection that should be applied (data classification). Administering database security, making emergency changes to data and migrating code to production are infrastructure tasks performed by custodians of the data.

S3-145   Which of the following is the **MOST** appropriate individual to ensure that new exposures have not been introduced into an existing application during the change management process?

A. System analyst
B. System user
C. Operations manager
D. Data security officer

B   System users, specifically the user acceptance testers, would be in the best position to note whether new exposures are introduced during the change management process. The system designer or system analyst, data security officer and operations manager would not be as closely involved in testing code changes.

S3-146   What is the **BEST** way to ensure users comply with organizational security requirements for password complexity?

A. Include password construction requirements in the security standards
B. Require each user to acknowledge the password requirements
C. Implement strict penalties for user noncompliance
D. Enable system-enforced password configuration

D   Automated controls are generally more effective in preventing improper actions. Policies and standards provide some deterrence, but are not as effective as automated controls.

S3-147   Which of the following is the **MOST** appropriate method for deploying operating system (OS) patches to production application servers?

A. Batch patches into frequent server updates
B. Initially load the patches on a test machine
C. Set up servers to automatically download patches
D. Automatically push all patches to the servers

B   Some patches can conflict with application code. For this reason, it is very important to first test all patches in a test environment to ensure that there are no conflicts with existing application systems. For this reason, choices C and D are incorrect as they advocate automatic updating. As for frequent server updates, this is an incomplete (vague) answer from the choices given.

S3-148   The **PRIMARY** reason for using metrics to evaluate information security is to:

A. identify security weaknesses.
B. justify budgetary expenditures.
C. enable steady improvement.
D. raise awareness on security issues.

C   The purpose of a metric is to facilitate and track continuous improvement. It will not permit the identification of all security weaknesses. It will raise awareness and help in justifying certain expenditures, but this is not its main purpose.

S3-149   What is the **BEST** method to confirm that all firewall rules and router configuration settings are adequate?

A. Periodic review of network configuration
B. Review intrusion detection system (IDS) logs for evidence of attacks
C. Periodically perform penetration tests
D. Daily review of server logs for evidence of hacker activity

C   Due to the complexity of firewall rules and router tables, plus the sheer size of intrusion detection systems (IDSs) and server logs, a physical review will be insufficient. The best approach for confirming the adequacy of these configuration settings is to periodically perform attack and penetration tests.

S3-150   Which of the following is **MOST** important for measuring the effectiveness of a security awareness program?

A. Reduced number of security violation reports
B. A quantitative evaluation to ensure user comprehension
C. Increased interest in focus groups on security issues
D. Increased number of security violation reports

B   To truly judge the effectiveness of security awareness training, some means of measurable testing is necessary to confirm user comprehension. Focus groups may or may not provide meaningful feedback but, in and of themselves, do not provide metrics. An increase or reduction in the number of violation reports may not be indicative of a high level of security awareness.

S3-151   Which of the following is the **MOST** important action to take when engaging third-party consultants to conduct an attack and penetration test?

A. Request a list of the software to be used
B. Provide clear directions to IT staff
C. Monitor intrusion detection system (IDS) and firewall logs closely
D. Establish clear rules of engagement

D   It is critical to establish a clear understanding on what is permissible during the engagement. Otherwise, the tester may inadvertently trigger a system outage or inadvertently corrupt files. Not as important, but still useful, is to request a list of what software will be used. As for monitoring the intrusion detection system (IDS) and firewall, and providing directions to IT staff, it is better not to alert those responsible for monitoring (other than at the management level), so that the effectiveness of that monitoring can be accurately assessed.

S3-152   Which of the following will **BEST** prevent an employee from using a USB drive to copy files from
          desktop computers?

          A.   Restrict the available drive allocation on all PCs
          B.   Disable universal serial bus (USB) ports on all desktop devices
          C.   Conduct frequent awareness training with noncompliance penalties
          D.   Establish strict access controls to sensitive information

A   Restricting the ability of a PC to allocate new drive letters ensures that universal serial bus (USB) drives
          or even CD-writers cannot be attached as they would not be recognized by the operating system. Disabling
          USB ports on all machines is not practical since mice and other peripherals depend on these connections.
          Awareness training and sanctions do not prevent copying of information nor do access controls.

S3-153   Good information security standards should:

          A.   define precise and unambiguous allowable limits.
          B.   describe the process for communicating violations.
          C.   address high-level objectives of the organization.
          D.   be updated frequently as new software is released.

A   A security standard should clearly state what is allowable; it should not change frequently. The process
          for communicating violations would be addressed by a security procedure, not a standard. High-level
          objectives of an organization would normally be addressed in a security policy.

S3-154   Good information security procedures should:

          A.   define the allowable limits of behavior.
          B.   underline the importance of security governance.
          C.   describe security baselines for each platform.
          D.   be updated frequently as new software is released.

D   Security procedures often have to change frequently to keep up with changes in software. Since a
          procedure is a how-to document, it must be kept up-to-date with frequent changes in software. A security
          standard—such as platform baselines—defines behavioral limits, not the how-to process; it should not
          change frequently. High-level objectives of an organization, such as security governance, would normally
          be addressed in a security policy.

S3-155   What is the **MAIN** drawback of e-mailing password-protected zip files across the Internet? They:

          A.   all use weak encryption.
          B.   are decrypted by the firewall.
          C.   may be quarantined by mail filters.
          D.   may be corrupted by the receiving mail server.

C   Often, mail filters will quarantine zip files that are password-protected since the filter (or the firewall) is
          unable to determine if the file contains malicious code. Many zip file products are capable of using strong
          encryption. Such files are not normally corrupted by the sending mail server.

S3-156   A major trading partner with access to the internal network is unwilling or unable to remediate serious information security exposures within its environment. Which of the following is the **BEST** recommendation?

A.   Sign a legal agreement assigning them all liability for any breach
B.   Remove all trading partner access until the situation improves
C.   Set up firewall rules restricting network traffic from that location
D.   Send periodic reminders advising them of their noncompliance

C   It is incumbent on an information security manager to see to the protection of their organization's network, but to do so in a manner that does not adversely affect the conduct of business. This can be accomplished by adding specific traffic restrictions for that particular location. Removing all access will likely result in lost business. Agreements and reminders do not protect the integrity of the network.

S3-157   Documented standards/procedures for the use of cryptography across the enterprise should **PRIMARILY**:

A.   define the circumstances where cryptography should be used.
B.   define cryptographic algorithms and key lengths.
C.   describe handling procedures of cryptographic keys.
D.   establish the use of cryptographic solutions.

A   There should be documented standards/procedures for the use of cryptography across the enterprise; they should define the circumstances where cryptography should be used. They should cover the selection of cryptographic algorithms and key lengths, but not define them precisely, and they should address the handling of cryptographic keys. However, this is secondary to how and when cryptography should be used. The use of cryptographic solutions should be addressed but, again, this is a secondary consideration.

S3-158   Which of the following is the **MOST** immediate consequence of failing to tune a newly installed intrusion detection system (IDS) with the threshold set to a low value?

A.   The number of false positives increases
B.   The number of false negatives increases
C.   Active probing is missed
D.   Attack profiles are ignored

A   Failure to tune an intrusion detection system (IDS) will result in many false positives, especially when the threshold is set to a low value. The other options are less likely given the fact that the threshold for sounding an alarm is set to a low value.

S3-159   What is the **MOST** appropriate change management procedure for the handling of emergency program changes?

A.   Formal documentation does not need to be completed before the change
B.   Business management approval must be obtained prior to the change
C.   Documentation is completed with approval soon after the change
D.   All changes must follow the same process

C   Even in the case of an emergency change, all change management procedure steps should be completed as in the case of normal changes. The difference lies in the timing of certain events. With an emergency change, it is permissible to obtain certain approvals and other documentation on "the morning after" once the emergency has been satisfactorily resolved. Obtaining business approval prior to the change is ideal but not always possible.

S3-160   Who is ultimately responsible for ensuring that information is categorized and that protective measures are taken?

     A. Information security officer
     B. Security steering committee
     C. Data owner
     D. Data custodian

B   Routine administration of all aspects of security is delegated, but senior management must retain overall responsibility. The information security officer supports and implements information security for senior management. The data owner is responsible for categorizing data security requirements. The data custodian supports and implements information security as directed.

S3-161   The **PRIMARY** focus of the change control process is to ensure that changes are:

     A. authorized.
     B. applied.
     C. documented.
     D. tested.

A   All steps in the change control process must be signed off on to ensure proper authorization. It is important that changes are applied, documented and tested; however, they are not the primary focus.

S3-162   An information security manager has been asked to develop a change control process. What is the **FIRST** thing the information security manager should do?

     A. Research best practices
     B. Meet with stakeholders
     C. Establish change control procedures
     D. Identify critical systems

B   No new process will be successful unless it is adhered to by all stakeholders; to the extent stakeholders have input, they can be expected to follow the process. Without consensus agreement from the stakeholders, the scope of the research is too wide; input on the current environment is necessary to focus research effectively. It is premature to implement procedures without stakeholder consensus and research. Without knowing what the process will be, the parameters to baseline are unknown as well.

S3-163   A critical device is delivered with a single user and password that is required to be shared for multiple users to access the device. An information security manager has been tasked with ensuring all access to the device is authorized. Which of the following would be the **MOST** efficient means to accomplish this?

     A. Enable access through a separate device that requires adequate authentication
     B. Implement manual procedures that require password change after each use
     C. Request the vendor to add multiple user IDs
     D. Analyze the logs to detect unauthorized access

A   Choice A is correct because it allows authentication tokens to be provisioned and terminated for individuals and also introduces the possibility of logging activity by individual. Choice B is not effective because users can circumvent the manual procedures. Choice C is not the best option because vendor enhancements may take time and development, and this is a critical device. Choice D could, in some cases, be an effective complementary control but, because it is detective, it would not be the most effective in this instance.

S3-164 Which of the following documents would be the **BEST** reference to determine whether access control mechanisms are appropriate for a critical application?

    A. User security procedures
    B. Business process flow
    C. IT security standards
    D. Regulatory requirements

C   IT management should ensure that mechanisms are implemented in line with IT security standards. Procedures are determined by the policy. A user security procedure does not describe the access control mechanism in place. The business process flow is not relevant to the access control mechanism. The organization's own policy and procedures should take into account regulatory requirements.

S3-165 Which of the following is the **MOST** important process that an information security manager needs to negotiate with an outsource service provider?

    A. The right to conduct independent security reviews
    B. A legally binding data protection agreement
    C. Encryption between the organization and the provider
    D. A joint risk assessment of the system

A   A key requirement of an outsource contract involving critical business systems is the establishment of the organization's right to conduct independent security reviews of the provider's security controls. A legally binding data protection agreement is also critical, but secondary to choice A, which permits examination of the actual security controls prevailing over the system and, as such, is the more effective risk management tool. Network encryption of the link between the organization and the provider may well be a requirement, but is not as critical since it would also be included in choice A. A joint risk assessment of the system in conjunction with the outsource provider may be a compromise solution, should the right to conduct independent security reviews of the controls related to the system prove contractually difficult.

S3-166 Which resource is the **MOST** effective in preventing physical access tailgating/piggybacking?

    A. Card key door locks
    B. Photo identification
    C. Awareness training
    D. Biometric scanners

C   Awareness training would most likely result in any attempted tailgating being challenged by the authorized employee. Choices A, B and D are physical controls that, by themselves, would not be effective against tailgating.

S3-167   In business critical applications, where shared access to elevated privileges by a small group is necessary, the **BEST** approach to implement adequate segregation of duties is to:

A. ensure access to individual functions can be granted to individual users only.
B. implement role-based access control in the application.
C. enforce manual procedures ensuring separation of conflicting duties.
D. create service accounts that can only be used by authorized team members.

B   Role-based access control is the best way to implement appropriate segregation of duties. Roles will have to be defined once and then the user could be changed from one role to another without redefining the content of the role each time. Access to individual functions will not ensure appropriate segregation of duties. Giving a user access to all functions and implementing, in parallel, a manual procedure ensuring segregation of duties is not an effective method, and would be difficult to enforce and monitor. Creating service accounts that can be used by authorized team members would not provide any help unless their roles are properly segregated.

S3-168   In business-critical applications, user access should be approved by the:

A. information security manager.
B. data owner.
C. data custodian.
D. business management.

B   A data owner is in the best position to validate access rights to users due to their deep understanding of business requirements and of functional implementation within the application. This responsibility should be enforced by the policy. An information security manager will coordinate and execute the implementation of the role-based access control. A data custodian will ensure that proper safeguards are in place to protect the data from unauthorized access; it is not the data custodian's responsibility to assign access rights. Business management is not, in all cases, the owner of the data.

S3-169   In organizations where availability is a primary concern, the **MOST** critical success factor of the patch management procedure would be the:

A. testing time window prior to deployment.
B. technical skills of the team responsible.
C. certification of validity for deployment.
D. automated deployment to all the servers.

A   Having the patch tested prior to implementation on critical systems is an absolute prerequisite where availability is a primary concern because deploying patches that could cause a system to fail could be worse than the vulnerability corrected by the patch. It makes no sense to deploy patches on every system. Vulnerable systems should be the only candidate for patching. Patching skills are not required since patches are more often applied via automated tools.

S3-170   To ensure that all information security procedures are functional and accurate, they should be designed with the involvement of:

       A.   end users.
       B.   legal counsel.
       C.   operational units.
       D.   audit management.

C   Procedures at the operational level must be developed by or with the involvement of operational units that will use them. This will ensure that they are functional and accurate. End users and legal counsel are normally not involved in procedure development. Audit management generally oversees information security operations but does not get involved at the procedural level.

S3-171   An information security manager reviewed the access control lists and observed that privileged access was granted to an entire department. Which of the following should the information security manager do **FIRST**?

       A.   Review the procedures for granting access
       B.   Establish procedures for granting emergency access
       C.   Meet with data owners to understand business needs
       D.   Redefine and implement proper access rights

C   An information security manager must understand the business needs that motivated the change prior to taking any unilateral action. Following this, all other choices could be correct depending on the priorities set by the business unit.

S3-172   A business partner of a factory has remote read-only access to material inventory to forecast future acquisition orders. An information security manager should **PRIMARILY** ensure that there is:

       A.   an effective control over connectivity and continuity.
       B.   a service level agreement (SLA) including code escrow.
       C.   a business impact analysis (BIA).
       D.   a third-party certification.

A   The principal risk focus is the connection procedures to maintain continuity in case of any contingency. Although an information security manager may be interested in the service level agreement (SLA), code escrow is not a concern. A business impact analysis (BIA) refers to contingency planning and not to system access. Third-party certification does not provide any assurance of controls over connectivity to maintain continuity.

S3-173   Which of the following should be in place before a black box penetration test begins?

       A.   IT management approval
       B.   Proper communication and awareness training
       C.   A clearly stated definition of scope
       D.   An incident response plan

C   Having a clearly stated definition of scope is most important to ensure a proper understanding of risk as well as success criteria. IT management approval may not be required based on senior management decisions. Communication, awareness and an incident response plan are not a necessary requirement. In fact, a penetration test could help promote the creation and execution of the incident response plan.

S3-174   What is the **MOST** important success factor in launching a corporate information security awareness program?

    A. Adequate budgetary support
    B. Centralized program management
    C. Top-down approach
    D. Experience of the awareness trainers

C   Senior management support will provide enough resources and will focus attention to the program; training should start at the top levels to gain support and sponsorship. Funding is not a primary concern. Centralized management does not provide sufficient support. Trainer experience, while important, is not the primary success factor.

S3-175   Which of the following events generally has the highest information security impact?

    A. Opening a new office
    B. Merging with another organization
    C. Relocating the data center
    D. Rewiring the network

B   Merging with or acquiring another organization causes a major impact on an information security management function because new vulnerabilities and risks are inherited. Opening a new office, moving the data center to a new site, or rewiring a network may have information security risks, but generally comply with corporate security policy and are easier to secure.

S3-176   The configuration management plan should be approved by:

    A. business process owners.
    B. the information security manager.
    C. the security steering committee.
    D. IT senior management.

D   Although business process owners, the information security manager and the security steering committee may provide input regarding a configuration management plan, its final approval is the primary responsibility of IT senior management.

S3-177   Who should determine the appropriate classification of accounting ledger data located on a database server and maintained by a database administrator in the IT department?

    A. Database administrator (DBA)
    B. Finance department management
    C. Information security manager
    D. IT department management

B   Data owners are responsible for determining data classification; in this case, management of the finance department would be the owners of accounting ledger data. The database administrator (DBA) and IT management are the custodians of the data who would apply the appropriate security levels for the classification, while the security manager would act as an advisor and enforcer.

S3-178   Which of the following is the **BEST** tool to maintain the currency and coverage of an information security program within an organization?

    A.   The program's governance oversight mechanisms
    B.   Information security periodicals and manuals
    C.   The program's security architecture and design
    D.   Training and certification of the information security team

A   While choices B, C and D will all assist the currency and coverage of the program, its governance oversight mechanisms are the best method.

S3-179   Which of the following would **BEST** assist an information security manager in measuring the existing level of development of security processes against their desired state?

    A.   Security audit reports
    B.   Balanced scorecard
    C.   Capability maturity model (CMM)
    D.   Systems and business security architecture

C   The capability maturity model (CMM) grades each defined area of security processes on a scale of 0 to 5 based on their maturity, and is commonly used by entities to measure their existing state and then determine the desired one. Security audit reports offer a limited view of the current state of security. Balanced scorecard is a document that enables management to measure the implementation of their strategy and assists in its translation into action. Systems and business security architecture explain the security architecture of an entity in terms of business strategy, objectives, relationships, risks, constraints and enablers, and provides a business-driven and business-focused view of security architecture.

S3-180   Who is responsible for raising awareness of the need for adequate funding to support risk mitigation plans?

    A.   Chief information officer (CIO)
    B.   Chief financial officer (CFO)
    C.   Information security manager
    D.   Business unit management

C   The information security manager is responsible for raising awareness of the need for adequate funding for risk-related mitigation plans. Even though the CIO, CFO and business unit management are involved in the final approval of fund expenditure, it is the information security manager who has the ultimate responsibility for raising awareness.

S3-181   Managing the life cycle of a digital certificate is a role of a(n):

    A.   system administrator.
    B.   security administrator.
    C.   system developer.
    D   independent trusted source.

D   Digital certificates must be managed by an independent trusted source in order to maintain trust in their authenticity. The other options are not necessarily entrusted with this capability.

S3-182   Change management procedures to ensure that disaster recovery/business continuity plans are kept up-to-date can be **BEST** achieved through which of the following?

    A. Reconciliation of the annual systems inventory to the disaster recovery/business continuity plans
    B. Periodic audits of the disaster recovery/business continuity plans
    C. Comprehensive walk-through testing
    D. Inclusion as a required step in the system life cycle process

D   Information security should be an integral component of the development cycle; thus, it should be included at the process level. Choices A, B and C are good mechanisms to ensure compliance, but would not be nearly as timely in ensuring that the plans are always up-to-date. Choice D is a preventive control, while choices A, B and C are detective controls.

S3-183   To reduce the possibility of service interruptions, an entity enters into contracts with multiple Internet service providers (ISPs). Which of the following would be the **MOST** important item to include?

    A. Service level agreements (SLAs)
    B. Right to audit clause
    C. Intrusion detection system (IDS) services
    D. Spam filtering services

A   Service level agreements (SLAs) will be most effective in ensuring that Internet service providers (ISPs) comply with expectations for service availability. Intrusion detection system (IDS) and spam filtering services would not mitigate (as directly) the potential for service interruptions. A right-to-audit clause would not be effective in mitigating the likelihood of a service interruption.

S3-184   To mitigate a situation where one of the programmers of an application requires access to production data, the information security manager could **BEST** recommend to:

    A. create a separate account for the programmer as a power user.
    B. log all of the programmers' activity for review by supervisor.
    C. have the programmer sign a letter accepting full responsibility.
    D. perform regular audits of the application.

B   It is not always possible to provide adequate segregation of duties between programming and operations in order to meet certain business requirements. A mitigating control is to record all of the programmers' actions for later review by their supervisor, which would reduce the likelihood of any inappropriate action on the part of the programmer. Choices A, C and D do not solve the problem.

S3-185   Before engaging outsourced providers, an information security manager should ensure that the organization's data classification requirements:

    A. are compatible with the provider's own classification.
    B. are communicated to the provider.
    C. exceed those of the outsourcer.
    D. are stated in the contract.

D   The most effective mechanism to ensure that the organization's security standards are met by a third party, would be a legal agreement. Choices A, B and C are acceptable options, but not as comprehensive or as binding as a legal contract.

S3-186   What is the **GREATEST** risk when there is an excessive number of firewall rules?

          A. One rule may override another rule in the chain and create a loophole
          B. Performance degradation of the whole network
          C. The firewall may not support the increasing number of rules due to limitations
          D. The firewall may show abnormal behavior and may crash or automatically shut down

A   If there are many firewall rules, there is a chance that a particular rule may allow an external connection although other associated rules are overridden. Due to the increasing number of rules, it becomes complex to test them and, over time, a loophole may occur.

S3-187   Which of the following would typically be the **MOST** effective physical security access control for the main entrance to a data center?

          A. Mantrap
          B. Biometric lock
          C. Closed-circuit television (CCTV)
          D. Security guard

B   A biometric device will ensure that only the authorized user can access the data center. A mantrap, by itself, would not be effective. Closed-circuit television (CCTV) and a security guard provide a detective control, but would not be as effective in authenticating the access rights of each individual.

S3-188   What is the **GREATEST** advantage of documented guidelines and operating procedures from a security perspective?

          A. Provide detailed instructions on how to carry out different types of tasks
          B. Ensure consistency of activities to provide a more stable environment
          C. Ensure compliance to security standards and regulatory requirements
          D. Ensure reusability to meet compliance to quality requirements

B   Developing procedures and guidelines to ensure that business processes address information security risk in a uniform manner is critical to the management of an information security program. Developing procedures and guidelines establishes a baseline for security program performance and consistency of security activities.

S3-189   What is the **BEST** way to ensure data protection upon termination of employment?

          A. Retrieve identification badge and card keys
          B. Retrieve all personal computer equipment
          C. Erase all of the employee's folders
          D. Ensure all logical access is removed

D   Ensuring all logical access is removed will guarantee that the former employee will not be able to access company data and that the employee's credentials will not be misused. Retrieving identification badge and card keys would only reduce the capability to enter the building. Retrieving the personal computer equipment and the employee's folders are necessary tasks, but that should be done as a second step.

S3-190   The **MOST** important reason for formally documenting security procedures is to ensure:

    A. processes are repeatable and sustainable.
    B. alignment with business objectives.
    C. auditability by regulatory agencies.
    D. objective criteria for the application of metrics.

A   Without formal documentation, it would be difficult to ensure that security processes are performed in the proper manner every time that they are performed. Alignment with business objectives is not a function of formally documenting security procedures. Processes should not be formally documented merely to satisfy an audit requirement. Although potentially useful in the development of metrics, creating formal documentation to assist in the creation of metrics is a secondary objective.

S3-191   Which of the following is the **BEST** approach for an organization desiring to protect its intellectual property?

    A. Conduct awareness sessions on intellectual property policy
    B. Require all employees to sign a nondisclosure agreement
    C. Promptly remove all access when an employee leaves the organization
    D. Restrict access to a need-to-know basis

D   Security awareness regarding intellectual property policy will not prevent violations of this policy. Requiring all employees to sign a nondisclosure agreement and promptly removing all access when an employee leaves the organization are good controls, but not as effective as restricting access to a need-to-know basis.

S3-192   The "separation of duties" principle is violated if which of the following individuals has update rights to the database access control list (ACL)?

    A. Data owner
    B. Data custodian
    C. Systems programmer
    D. Security administrator

C   A systems programmer should not have privileges to modify the access control list (ACL) because this would give the programmer unlimited control over the system. The data owner would request and approve updates to the ACL, but it is not a violation of the separation of duties principle if the data owner has update rights to the ACL. The data custodian and the security administrator could carry out the updates on the ACL since it is part of their duties as delegated to them by the data owner.

S3-193   An account with full administrative privileges over a production file is found to be accessible by a member of the software development team. This account was set up to allow the developer to download nonsensitive production data for software testing purposes. Assuming all options are possible, which of the following should the information security manager recommend?

    A.   Restrict account access to read only
    B.   Log all usage of this account
    C.   Suspend the account and activate only when needed
    D.   Require that a change request be submitted for each download

A   Administrative accounts have permission to change data. This is not required for the developers to perform their tasks. Unauthorized change will damage the integrity of the data. Logging all usage of the account, suspending the account and activating only when needed, and requiring that a change request be submitted for each download will not reduce the exposure created by this excessive level of access. Restricting the account to read only access will ensure that file integrity can be maintained while permitting access.

S3-194   Which of the following is the **BEST** indicator that security controls are performing effectively?

    A.   The monthly service level statistics indicate minimal impact from security issues.
    B.   The cost of implementing security controls is less than the value of the assets.
    C.   The percentage of systems that are compliant with security standards is satisfactory.
    D.   Audit reports do not reflect any significant findings on security.

A   The best indicator of effective security control is the evidence of little disruption to business operations. Choices B, C and D can support this evidence, but are supplemental to choice A.

S3-195   An organization's information security manager has been asked to hire a consultant to help assess the maturity level of the organization's information security management. The **MOST** important element of the request for proposal (RFP) is the:

    A.   references from other organizations.
    B.   past experience of the engagement team.
    C.   sample deliverable.
    D.   methodology to be used in the assessment.

D   Methodology illustrates the process and formulates the basis to align expectations and the execution of the assessment. This also provides a picture of what is required of all parties involved in the assessment. References from other organizations are important, but not as important as the methodology used in the assessment. Past experience of the engagement team is not as important as the methodology used. Sample deliverables only tell how the assessment is presented, not the process.

S3-196   Several business units reported problems with their systems after multiple security patches were deployed. The **FIRST** step in handling this problem would be to:

    A.   assess the problems and institute rollback procedures, if needed.
    B.   disconnect the systems from the network until the problems are corrected.
    C.   immediately uninstall the patches from these systems.
    D.   immediately contact the vendor regarding the problems that occurred.

A   Assessing the problems and instituting rollback procedures as needed would be the best course of action. Choices B and C would not identify where the problem was, and may in fact make the problem worse. Choice D is part of the assessment.

S3-197   When defining a service level agreement (SLA) regarding the level of data confidentiality that is handled by a third-party service provider, the **BEST** indicator of compliance would be the:

    A. access control matrix.
    B. encryption strength.
    C. authentication mechanism.
    D. data repository.

A   The access control matrix is the best indicator of the level of compliance with the service level agreement (SLA) data confidentiality clauses. Encryption strength, authentication mechanism and data repository might be defined in the SLA but are not confidentiality compliance indicators.

S3-198   The **PRIMARY** reason for involving information security at each stage in the systems development life cycle (SDLC) is to identify the security implications and potential solutions required for:

    A. identifying vulnerabilities in the system.
    B. sustaining the organization's security posture.
    C. the existing systems that will be affected.
    D. complying with segregation of duties.

B   It is important to maintain the organization's security posture at all times. The focus should not be confined to the new system being developed or acquired, or to the existing systems in use. Segregation of duties is only part of a solution to improving the security of the systems, not the primary reason to involve security in the systems development life cycle (SDLC).

S3-199   The implementation of continuous monitoring controls is the **BEST** option where:

    A. incidents may have a high impact and frequency
    B. legislation requires strong information security controls
    C. incidents may have a high impact but low frequency
    D. electronic commerce is a primary business driver

A   Continuous monitoring control initiatives are expensive, so they have to be used in areas where the risk is at its greatest level. These areas are the ones with high impact and high frequency of occurrence. Regulations and legislations that require tight IT security measures focus on requiring organizations to establish an IT security governance structure that manages IT security with a risk-based approach, so each organization decides which kinds of controls are implemented. Continuous monitoring is not necessarily a requirement. Measures such as contingency planning are commonly used when incidents rarely happen but have a high impact each time they happen. Continuous monitoring is unlikely to be necessary. Continuous control monitoring initiatives are not needed in all electronic commerce environments. There are some electronic commerce environments where the impact of incidents is not high enough to support the implementation of this kind of initiative.

S3-200   A third party was engaged to develop a business application. Which of the following is the **BEST** test for the existence of back doors?

A.   System monitoring for traffic on network ports
B.   Security code reviews for the entire application
C.   Reverse engineering the application binaries
D.   Running the application from a high-privileged account on a test system

B   Security code reviews for the entire application is the best measure and will involve reviewing the entire source code to detect all instances of back doors. System monitoring for traffic on network ports would not be able to detect all instances of back doors and is time consuming and would take a lot of effort. Reverse engineering the application binaries may not provide any definite clues. Back doors will not surface by running the application on high-privileged accounts since back doors are usually hidden accounts in the applications.

S3-201   An information security manager reviewing firewall rules will be **MOST** concerned if the firewall allows:

A.   source routing.
B.   broadcast propagation.
C.   unregistered ports.
D.   nonstandard protocols.

A   If the firewall allows source routing, any outsider can carry out spoofing attacks by stealing the internal (private) IP addresses of the organization. Broadcast propagation, unregistered ports and nonstandard protocols do not create a significant security exposure.

S3-202   What is the **MOST** cost-effective means of improving security awareness of staff personnel?

A.   Employee monetary incentives
B.   User education and training
C.   A zero-tolerance security policy
D.   Reporting of security infractions

B   User education and training is the most cost-effective means of influencing staff to improve security since personnel are the weakest link in security. Incentives perform poorly without user education and training. A zero-tolerance security policy would not be as good as education and training. Users would not have the knowledge to accurately interpret and report violations without user education and training.

S3-203   Which of the following is the **MOST** effective at preventing an unauthorized individual from following an authorized person through a secured entrance (tailgating or piggybacking)?

A.   Card-key door locks
B.   Photo identification
C.   Biometric scanners
D.   Awareness training

D   Awareness training would most likely result in any attempted tailgating being challenged by the authorized employee. The other choices are physical controls which by themselves would not be effective against tailgating.

S3-204   Data owners will determine what access and authorizations users will have by:

A. delegating authority to data custodian.
B. cloning existing user accounts.
C. determining hierarchical preferences.
D. mapping to business needs.

D   Access and authorizations should be based on business needs. Data custodians implement the decisions
          made by data owners. Access and authorizations are not to be assigned by cloning existing user accounts
          or determining hierarchical preferences. By cloning, users may obtain more access rights and privileges
          than is required to do their job. Hierarchical preferences may be based on individual preferences and not on
          business needs.

S3-205   Which of the following is the **MOST** likely outcome of a well-designed information security
          awareness course?

A. Increased reporting of security incidents to the incident response function
B. Decreased reporting of security incidents to the incident response function
C. Decrease in the number of password resets
D. Increase in the number of identified system vulnerabilities

A   A well-organized information security awareness course informs all employees of existing security policies, the
          importance of following safe practices for data security and the need to report any possible security incidents to
          the appropriate individuals in the organization. The other choices would not be the likely outcomes.

S3-206   Which item would be the **BEST** to include in the information security awareness training program for new
          general staff employees?

A. Review of various security models
B. Discussion of how to construct strong passwords
C. Review of roles that have privileged access
D. Discussion of vulnerability assessment results

B   All new employees will need to understand techniques for the construction of strong passwords. The other
          choices would not be applicable to general staff employees.

S3-207   A critical component of a continuous improvement program for information security is:

A. program metrics.
B. developing a service level agreement (SLA) for security.
C. tying corporate security standards to a recognized international standard.
D. ensuring regulatory compliance.

A   If an organization is unable to take measurements over time that provide data regarding key aspects of its
          security program, then continuous improvement is not possible. Although desirable, developing a service level
          agreement (SLA) for security, tying corporate security standards to a recognized international standard and
          ensuring regulatory compliance are not critical components for a continuous improvement program.

S3-208   An organization has implemented an enterprise resource planning (ERP) system used by 500 employees
           from various departments. Which of the following access control approaches is **MOST** appropriate?

           A. Rule-based
           B. Mandatory
           C. Discretionary
           D. Role-based

D   Role-based access control is effective and efficient in large user communities because it controls system
           access by the roles defined for groups of users. Users are assigned to the various roles and the system
           controls the access based on those roles. Rule-based access control needs to define the access rules, which
           is troublesome and error prone in large organizations. In mandatory access control, the individual's access
           to information resources needs to be defined, which is troublesome in large organizations. In discretionary
           access control, users have access to resources based on predefined sets of principles, which is an inherently
           insecure approach.

S3-209   An organization plans to contract with an outside service provider to host its corporate web site. The
           MOST important concern for the information security manager is to ensure that:

           A. an audit of the service provider uncovers no significant weakness.
           B. the contract includes a nondisclosure agreement (NDA) to protect the organization's intellectual property.
           C. the contract should mandate that the service provider will comply with security policies.
           D. the third-party service provider conducts regular penetration testing.

C   It is critical to include the security requirements in the contract based on the company's security policy to
           ensure that the necessary security controls are implemented by the service provider. The audit is normally
           a one-time effort and cannot provide ongoing assurance of the security. A nondisclosure agreement (NDA)
           should be part of the contract; however, it is not critical to the security of the web site. Penetration testing
           alone would not provide total security to the web site; there are lots of controls that cannot be tested
           through penetration testing.

S3-210   Which of the following is the **MAIN** objective in contracting with an external company to perform
           penetration testing?

           A. To mitigate technical risks
           B. To have an independent certification of network security
           C. To receive an independent view of security exposures
           D. To identify a complete list of vulnerabilities

C   Even though the organization may have the capability to perform penetration testing with internal
           resources, third-party penetration testing should be performed to gain an independent view of the security
           exposure. Mitigating technical risks is not a direct result of a penetration test. A penetration test would not
           provide certification of network security nor provide a complete list of vulnerabilities.

S3-211   An organization plans to outsource its customer relationship management (CRM) to a third-party service provider. Which of the following should the organization do **FIRST**?

    A. Request that the third-party provider perform background checks on their employees.
    B. Perform an internal risk assessment to determine needed controls.
    C. Audit the third-party provider to evaluate their security controls.
    D. Perform a security assessment to detect security vulnerabilities.

B   An internal risk assessment should be performed to identify the risk and determine needed controls. A background check should be a standard requirement for the service provider. Audit objectives should be determined from the risk assessment results. Security assessment does not cover the operational risks.

S3-212   Which of the following would raise security awareness among an organization's employees?

    A. Distributing industry statistics about security incidents
    B. Monitoring the magnitude of incidents
    C. Encouraging employees to behave in a more conscious manner
    D. Continually reinforcing the security policy

D   Employees must be continually made aware of the policy and expectations of their behavior. Choice A would have little relevant bearing on the employee's behavior. Choice B does not involve the employees. Choice C could be an aspect of continual reinforcement of the security policy.

S3-213   Which of the following is the **MOST** appropriate method of ensuring password strength in a large organization?

    A. Attempt to reset several passwords to weaker values
    B. Install code to capture passwords for periodic audit
    C. Sample a subset of users and request their passwords for review
    D. Install strong password settings on each platform

D   Adjusting password settings on each platform will be the most efficient method for ensuring password strength while not compromising the integrity of the passwords. Attempting to reset several passwords to weaker values may not highlight certain weaknesses. Installing code to capture passwords for periodic audit, and sampling a subset of users and requesting their passwords for review, would compromise the integrity of the passwords.

S3-214   Which of the following is the **BEST** approach for improving information security management processes?

    A. Conduct periodic security audits.
    B. Perform periodic penetration testing.
    C. Define and monitor security metrics.
    D. Survey business units for feedback.

C   Defining and monitoring security metrics is a good approach to analyze the performance of the security management process since it determines the baseline and evaluates the performance against the baseline to identify an opportunity for improvement. This is a systematic and structured approach to process improvement. Audits will identify deficiencies in established controls; however, they are not effective in evaluating the overall performance for improvement. Penetration testing will only uncover technical vulnerabilities, and cannot provide a holistic picture of information security management. Feedback is subjective and not necessarily reflective of true performance.

S3-215   When developing metrics to measure and monitor information security programs, the information security manager should ensure that the metrics reflect the:

A. residual risks.
B. levels of security.
C. security objectives.
D. statistics of security incidents.

C   Metrics should be developed based on security objectives, so they can measure the effectiveness and efficiency of information security controls. Metrics are not only used to measure the results of the security controls (residual risks and levels of security), but also the attributes of the control implementation. Not only statistics are collected, but other attributes of the information security controls should also be considered.

S3-216   An effective way of protecting applications against Structured Query Language (SQL) injection vulnerability is to:

A. validate and sanitize client side inputs.
B. harden the database listener component.
C. normalize the database schema to the third normal form.
D. ensure that the security patches are updated on operating systems.

A   SQL injection vulnerability arises when crafted or malformed user inputs are substituted directly in SQL queries, resulting in information leakage. Hardening the database listener does enhance the security of the database; however, it is unrelated to the SQL injection vulnerability. Normalization is related to the effectiveness and efficiency of the database but not to SQL injection vulnerability. SQL injections may also be observed in normalized databases. SQL injection vulnerability exploits the SQL query design, not the operating system.

S3-217   The root cause of a successful cross site request forgery (XSRF) attack against an application is that the vulnerable application:

A. uses multiple redirects for completing a data commit transaction.
B. has implemented cookies as the sole authentication mechanism.
C. has been installed with a non-legitimate license key.
D. is hosted on a server along with other applications.

B   XSRF exploits inadequate authentication mechanisms in web applications that rely only on elements such as cookies when performing a transaction. XSRF is related to an authentication mechanism, not to redirection. Option C is related to intellectual property rights, not to XSRF vulnerability. Merely hosting multiple applications on the same server is not the root cause of this vulnerability.

S3-218   An organization is entering into an agreement with a new business partner to conduct customer mailings. What is the **MOST** important action that the information security manager needs to perform?

A. A due diligence security review of the business partner's security controls
B. Ensuring that the business partner has an effective business continuity program
C. Ensuring that the third party is contractually obligated to all relevant security requirements
D. Talking to other clients of the business partner to check references for performance

C   The key requirement is that the information security manager ensures that the third party is contractually bound to follow the appropriate security requirements for the process being outsourced. This protects both organizations. All other steps are contributory to the contractual agreement, but are not key.

S3-219   An organization that outsourced its payroll processing performed an independent assessment of the
          security controls of the third party, per policy requirements. Which of the following is the **MOST** useful
          requirement to include in the contract?

          A. Right to audit
          B. Nondisclosure agreement
          C. Proper firewall implementation
          D. Dedicated security manager for monitoring compliance

A   Right to audit would be the most useful requirement since this would provide the company the ability to
          perform a security audit/assessment whenever there is a business need to examine whether the controls are
          working effectively at the third party. Choices B, C and D are important requirements and can be examined
          during the audit. A dedicated security manager would be a costly solution and not always feasible for
          most situations.

S3-220   Which of the following is the **MOST** critical activity to ensure the ongoing security of outsourced IT services?

          A. Provide security awareness training to the third-party provider's employees
          B. Conduct regular security reviews of the third-party provider
          C. Include security requirements in the service contract
          D. Request that the third-party provider comply with the organization's information security policy

B   Regular security audits and reviews of the practices of the provider to prevent potential information security
          damage will help verify the security of outsourced services. Depending on the type of services outsourced,
          security awareness may not be necessary. Security requirements should be included in the contract, but
          what is most important is verifying that the requirements are met by the provider. It is not necessary to
          require the provider to fully comply with the policy if only some of the policy is related and applicable.

S3-221   An organization's operations staff places payment files in a shared network folder and then the
          disbursement staff picks up the files for payment processing. This manual intervention will be automated
          some months later, thus cost-efficient controls are sought to protect against file alterations. Which of the
          following would be the **BEST** solution?

          A. Design a training program for the staff involved to heighten information security awareness
          B. Set role-based access permissions on the shared folder
          C. The end user develops a PC macro program to compare sender and recipient file contents
          D. Shared folder operators sign an agreement to pledge not to commit fraudulent activities

B   Ideally, requesting that the IT department develop an automated integrity check would be desirable, but
          given the temporary nature of the problem, the risk can be mitigated by setting stringent access permissions
          on the shared folder. Operations staff should only have write access and disbursement staff should only
          have read access, and everyone else, including the administrator, should be disallowed. An information
          security awareness program and/or signing an agreement to not engage in fraudulent activities may help
          deter attempts made by employees; however, as long as employees see a chance of personal gain when
          internal control is loose, they may embark on unlawful activities such as alteration of payment files. A
          PC macro would be an inexpensive automated solution to develop with control reports. However, sound
          independence or segregation of duties cannot be expected in the reconciliation process since it is run by an
          end-user group. Therefore, this option may not provide sufficient proof.

S3-222   Which of the following **BEST** ensures that security risks will be reevaluated when modifications in application developments are made?

    A. A problem management process
    B. Background screening
    C. A change control process
    D. Business impact analysis (BIA)

C   A change control process is the methodology that ensures that anything that could be impacted by a development change will be reevaluated. Problem management is the general process intended to manage all problems, not those specifically related to security. Background screening is the process to evaluate employee references when they are hired. BIA is the methodology used to evaluate risks in the business continuity process.

S3-223   In which of the following system development life cycle (SDLC) phases are access control and encryption algorithms chosen?

    A. Procedural design
    B. Architectural design
    C. System design specifications
    D. Software development

C   The system design specifications phase is when security specifications are identified. The procedural design converts structural components into a procedural description of the software. The architectural design is the phase that identifies the overall system design, but not the specifics. Software development is too late a stage since this is the phase when the system is already being coded.

S3-224   Which of the following is generally considered a fundamental component of an information security program?

    A. Role-based access control systems
    B. Automated access provisioning
    C. Security awareness training
    D. Intrusion prevention systems (IPSs)

C   Without security awareness training, many components of the security program may not be effectively implemented. The other options may or may not be necessary, but are discretionary.

S3-225   How would an organization know if its new information security program is accomplishing its goals?

    A. Key metrics indicate a reduction in incident impacts.
    B. Senior management has approved the program and is supportive of it.
    C. Employees are receptive to changes that were implemented.
    D. There is an immediate reduction in reported incidents.

A   Choice A is correct since an effective security program will show a trend in impact reduction. Choices B and C may well derive from a performing program, but are not as significant as Choice A. Choice D may indicate that it is not successful.

S3-226   A benefit of using a full disclosure (white box) approach as compared to a blind (black box) approach to penetration testing is that:

A. it simulates the real-life situation of an external security attack.
B. human intervention is not required for this type of test.
C. less time is spent on reconnaissance and information gathering.
D. critical infrastructure information is not revealed to the tester.

C   Data and information required for penetration are shared with the testers, thus eliminating time that would otherwise have been spent on reconnaissance and gathering of information. Blind (black box) penetration testing is closer to real life than full disclosure (white box) testing. There is no evidence to support that human intervention is not required for this type of test. A full disclosure (white box) methodology requires the knowledge of the subject being tested.

S3-227   Which of the following is the **BEST** method to reduce the number of incidents of employees forwarding spam and chain e-mail messages?

A. Acceptable use policy
B. Setting low mailbox limits
C. User awareness training
D. Taking disciplinary action

C   User awareness training would help in reducing the incidents of employees forwarding spam and chain e-mails since users would understand the risks of doing so and the impact on the organization's information system. An acceptable use policy, signed by employees, would legally address the requirements but merely having a policy is not the best measure. Setting low mailbox limits and taking disciplinary action are a reactive approach and may not help in obtaining proper support from employees.

S3-228   Which of the following is the **BEST** approach to mitigate online brute-force attacks on user accounts?

A. Passwords stored in encrypted form
B. User awareness
C. Strong passwords that are changed periodically
D. Implementation of lock-out policies

D   Implementation of account lock-out policies significantly inhibits brute-force attacks. In cases where this is not possible, strong passwords that are changed periodically would be an appropriate choice. Passwords stored in encrypted form will not defeat an online brute-force attack if the password itself is easily guessed. User awareness would help but is not the best approach of the options given.

S3-229   Which of the following measures is the **MOST** effective deterrent against disgruntled staff abusing their privileges?

A. Layered defense strategy
B. System audit log monitoring
C. Signed acceptable use policy
D. High-availability systems

C   A layered defense strategy would only prevent those activities that are outside of the user's privileges. A signed acceptable use policy is often an effective deterrent against malicious activities because of the potential for termination of employment and/or legal actions being taken against the individual. System audit log monitoring is after the fact and may not be effective. High-availability systems have high costs and are not always feasible for all devices and components or systems.

S3-230   The advantage of sending messages using steganographic techniques, as opposed to utilizing encryption, is that:

    A.   the existence of messages is unknown.
    B.   required key sizes are smaller.
    C.   traffic cannot be sniffed.
    D.   reliability of the data is higher in transit.

A   The existence of messages is hidden when using steganography. This is the greatest risk. Keys are relevant for encryption and not for steganography. Sniffing of steganographic traffic is also possible. Option D is not relevant.

S3-231   As an organization grows, exceptions to information security policies that were not originally specified may become necessary at a later date. In order to ensure effective management of business risks, exceptions to such policies should be:

    A.   considered at the discretion of the information owner.
    B.   approved by the next higher person in the organizational structure.
    C.   formally managed within the information security management framework.
    D.   reviewed and approved by the security manager.

C   A formal process for managing exceptions to information security policies and standards should be included as part of the information security management framework. The other options may be contributors to the process but do not in themselves constitute a formal process.

S3-232   There is reason to believe that a recently modified web application has allowed unauthorized access. Which is the **BEST** way to identify an application backdoor?

    A.   Black box pen test
    B.   Security audit
    C.   Source code review
    D.   Vulnerability scan

C   Source code review is the best way to find and remove an application backdoor. Application backdoors can be almost impossible to identify using a black box pen test or a security audit. A vulnerability scan will only find "known" vulnerability patterns and will therefore not find a programmer's application backdoor.

S3-233   Simple Network Management Protocol v2 (SNMP v2) is used frequently to monitor networks. Which of the following vulnerabilities does it always introduce?

    A.   Remote buffer overflow
    B.   Cross site scripting
    C.   Clear text authentication
    D.   Man-in-the-middle attack

C   One of the main problems with using SNMP v1 and v2 is the clear text "community string" that it uses to authenticate. It is easy to sniff and reuse. Most times, the SNMP community string is shared throughout the organization's servers and routers, making this authentication problem a serious threat to security. There have been some isolated cases of remote buffer overflows against SNMP daemons, but generally that is not a problem. Cross site scripting is a web application vulnerability that is not related to SNMP. A man-in-the-middle attack against a user datagram protocol (UDP) makes no sense since there is no active session; every request has the community string and is answered independently.

S3-234   Which of the following is the **FIRST** phase in which security should be addressed in the development cycle of a project?

    A. Design
    B. Implementation
    C. Application security testing
    D. Feasibility

D   Information security should be considered at the earliest possible stage. Security requirements must be defined before you enter into design specification, although changes in design may alter these requirements later on. Security requirements defined during system implementation are typically costly add-ons that are frequently ineffective. Application security testing occurs after security has been implemented.

S3-235   Which web application attack facilitates unauthorized access to a database?

    A. Cross site request forgery
    B. Structured Query Language (SQL) injection
    C. Metasploit
    D. Cross site scripting

B   SQL injection is a vulnerability that enables an attacker to execute commands through the web application, directly into the database. By accessing the database, data can potentially be read and altered. Cross site request forgery and cross site scripting attacks occur in the victim's web browser and have no access to database data. Metasploit is an exploit development suite that could allow access to a database by using one of its buffer overflow attacks, but this would not be a web application layer attack.

S3-236   Which of the following remote administration protocols is **MOST** secure?

    A. Secure Shell (SSH)
    B. Virtual Network Computing (VNC)
    C. Telnet
    D. HTTP

A   SSH replaced Telnet as a secure alternative. Today SSH v2.0 is one of the most secure remote administration protocols available. It can prevent man-in-the-middle attacks, sniffing and spoofing. Telnet and HTTP are clear text protocols and are therefore high risk. VNC has multiple vulnerabilities, including weak authentication and password storage, as well as frequently unencrypted connections.

S3-237   At what point should a risk assessment of a new process occur to determine appropriate controls?
It should occur:

    A. only at the beginning and at the end of the new process.
    B. during the entire life cycle of the process.
    C. at the appropriate point since timing of assessments will differ for processes.
    D. depending upon laws and regulations.

B   A risk assessment should be conducted during the entire life cycle of a new or a changed process. This allows an understanding of how implementation of an early control will affect control needs later on in a process.

S3-238   What would be the **MOST** significant security risk when using wireless local area network (LAN) technology?

A. Man-in-the-middle attack
B. Spoofing of data packets
C. Rogue access point
D. Session hijacking

C   A rogue access point masquerades as a legitimate access point. The risk is that legitimate users may connect through this access point and have their traffic monitored. All other choices are not dependent on the use of a wireless local area network (LAN) technology.

S3-239   Which of the following will **BEST** prevent external security attacks?

A. Static IP addressing
B. Network address translation
C. Background checks for temporary employees
D. Securing and analyzing system access logs

B   Network address translation is helpful by having internal addresses that are nonroutable. Background checks of temporary employees are more likely to prevent an attack launched from within the enterprise. Static IP addressing does little to prevent an attack. Writing all computer logs to removable media does not help in preventing an attack.

S3-240   To determine the selection of controls required to meet business objectives, an information security manager should:

A. prioritize the use of role-based access controls.
B. focus on key controls.
C. restrict controls to only critical applications.
D. focus on automated controls.

B   Key controls primarily reduce risk and are most effective for the protection of information assets. The other choices could be examples of possible key controls.

S3-241   The purpose of a corrective control is to:

A. reduce adverse events.
B. indicate compromise.
C. mitigate impact.
D. ensure compliance.

C   Corrective controls serve to reduce or mitigate impacts, such as providing recovery capabilities. Preventive controls reduce adverse events, such as firewalls. Compromise can be detected by detective controls, such as intrusion detection systems (IDSs). Compliance could be ensured by preventive controls, such as access controls.

S3-242   An enterprise requires the use of Windows XP Service Pack 3 version on all desktops and Windows 2003 Service Pack 1 version on all servers. This is an example of a:

    A. policy.
    B. guideline.
    C. standard.
    D. procedure.

C   A standard includes required hardware and software mechanisms without describing the settings used in that software. Required operating system software is an example of a standard. A standard sets the minimum requirements for required software or hardware. A guideline is not as mandatory as a standard requirement and is more like a recommendation. Procedures are usually detailed, step-by-step required actions.

S3-243   Which of the following would **BEST** help to uncover points of buffer overflow in an application?

    A. Integration testing
    B. A data flow diagram
    C. A code review
    D. Application logs

C   Buffer overflow in an application is primarily due to improper or inadequate specification of input length and this can be best uncovered through code reviews.

S3-244   Which of the following would be the **BEST** way to improve employee attitude toward and commitment to information security?

    A. Implement restrictive controls.
    B. Customize methods training to the audience.
    C. Apply administrative penalties.
    D. Initiate stronger supervision.

B   Cultural differences will dictate the best behavior modification techniques. For example, some cultures value relationships over monetary rewards. The other choices may work in certain circumstances, enterprises and geographic locations, but not in others.

S3-245   When considering outsourcing services, at what point should information security become involved in the vendor management process?

    A. During contract negotiation
    B. Upon request for assistance from the business unit
    C. When requirements are being established
    D. When a security incident occurs

C   Information security should be involved in the vendor or third-party management process from the beginning of the selection process, when the business is defining what it needs. This will ensure that all bids for the service take into consideration, and reflect in bid prices, the security requirements. Waiting until later in the process can lead to vendors having to re-bid and can disrupt negotiations. Waiting until after the contract is signed can expose the enterprise to significant security risk, with little recourse to correct, because the contract has already been executed. There may be situations where information security involvement is not required, but those situations would be established by conducting an initial risk assessment.

S3-246 Which of the following should be performed **EXCLUSIVELY** by the information security department?

    A. Monitoring unauthorized access to operating systems
    B. Configuring user access to operating systems
    C. Approving operating system access standards
    D. Configuring the firewall to protect operating systems

C The approval of standards to meet the requirements of policies should be performed by the information security department. Separation of duties will be required to ensure that operational constraints do not result in standards not being met. The implementation of the standards may be performed in conjunction with the IT department. The other functions may or may not be performed by the information security department. Approving security standards is performed exclusively by the information security department.

S3-247 Which of the following is the **MOST** critical success factor of an information security program?

    A. Developing information security policies and procedures
    B. Senior management commitment
    C. Conducting security training and awareness for all users
    D. Establishing an information security management system

B Without senior management commitment, it would be difficult to implement a successful information security program. The other choices are valuable, but not the most critical.

S3-248 The development of an information security program begins with:

    A. a comprehensive risk assessment and analysis.
    B. the development of a security architecture.
    C. completion of a controls statement of applicability.
    D. an effective information security strategy.

D The process of developing information security governance structures, achieving organizational adoption, and developing an implementation strategy will define the scope and responsibilities of the security program. Assessing and analyzing risk is required to develop a strategy and will provide some of the information needed to develop the strategy, but will not define the scope and charter of the security program. A security architecture is a part of implementation subsequent to developing the strategy. The applicability statement is a part of strategy implementation using ISO 27001 or 27002 subsequent to determining the scope and responsibilities of the program.

S3-249 Which of the following constitutes the **MAIN** project activities undertaken in developing an information security program?

    A. Controls design and deployment
    B. Security organization development
    C. Logical and conceptual architecture design
    D. Development of risk management objectives

A The majority of program development activities will involve designing, testing and deploying controls that achieve the risk management objectives. The security organization should be fairly well developed prior to attempting to implement a security program. Conceptual and logical architecture designs should have been completed as a part of strategy and road-map development. Risk management objectives are part of strategy development.

S3-250   A financial institution plans to allocate information security resources to each of its business divisions. The priority of focus in security activity should be in areas:

    A.   where strict regulatory requirements apply.
    B.   that require the shortest recovery time objective (RTO).
    C.   that can maximize return on security investment (ROSI).
    D.   where threat likelihood and impact are greatest.

D   While regulatory requirements may be a major consideration, there may be other areas of greater threat and impact to the enterprise. Watching the RTO requirement is very important from a business continuity perspective, but this only illustrates a part of the information security framework. Regulatory compliance may also touch upon RTO initiatives. It is difficult to set up a single formula so that the most profitable business line always has the most critical information security initiatives in the enterprise. Therefore, this is not always a good choice.

S3-251   Which of the following should be responsible for final approval of security patch implementation?

    A.   The application development manager
    B.   The business asset owner
    C.   The information security officer
    D.   The business continuity coordinator

B   In order to ensure that no serious business interruption takes place due to any unexpected problems, it is important to bring business asset owners into the final sign-off loop when a security patch is being released. When business logic has been modified, the application development team may be involved in testing; however, the team's involvement will be less necessary when a security patch is being released. The information security officer is informed of security patches currently being released; however, the information security officer's approval may not always be required for patch release. A business continuity coordinator would not be involved in approving security patches in normal day-to-day operations.

S3-252   Which is the **FIRST** thing that should be determined by the information security manager when developing an information security program?

    A.   The control objectives
    B.   The strategic aims
    C.   The desired outcomes
    D.   The logical architecture

C   Without determining the desired outcomes of the security program, it will be difficult or impossible to determine a viable strategy, control objectives and logical architecture.

S3-253   Which of the following is the **BEST** way to mitigate the risk of the database administrator reading sensitive data from the database?

    A.   Log all access to sensitive data.
    B.   Employ application-level encryption.
    C.   Install a database monitoring solution.
    D.   Develop a data security policy.

B   Data encrypted at the application level that is stored in a database cannot be viewed in clear text, even by the database administrator (DBA). Access logging can be easily turned off by the DBA. A database monitoring solution can be bypassed by the DBA. A security policy will only be effective if the DBA chooses to adhere to the policy.

S3-254   In a large enterprise, an information security awareness program will be **MOST** effective if it is:

    A. developed by a professional training company.
    B. embedded into the orientation process.
    C. customized to the audience using the appropriate delivery channel.
    D. required by the information security policy.

C   An awareness program should be customized for different types of audiences, e.g., for new employees, system administration, sales and delivery channels such as posters or e-learning. It does not have to be developed by a professional training company to make it effective. The awareness program should be embedded into the orientation process for new employees. Being required by policy does not make the program more effective.

S3-255   The prioritization of security spending and budgeting would **PRIMARILY** be based on:

    A. identified levels of risk.
    B. industry trends.
    C. an increased cost of services.
    D. the allocated revenue of the enterprise.

A   The first required action is to conduct a risk assessment of the enterprise's key processes to identify control gaps and determine where investments should be made to mitigate risks and to determine order of prioritization. This must be conducted with consideration of enterprise goals and strategy. Prioritization should not be based on the trends at other organizations since each organization has unique requirements and business objectives. Prioritization by cost alone is not aligned with a risk-based approach. Although the revenue may increase, it is not wise to link the IT budget to a fixed percentage of revenue since this could lead to spending more than is necessary to effectively address risk.

S3-256   The data backup policy will contain which of the following?

    A. Criteria for data backup
    B. Personnel responsible for backup
    C. A data backup schedule
    D. A list of systems to be backed up

A   A policy is a high-level management intent and will essentially contain the criteria to be followed for backing up any data such as critical data, confidential data and project data, and the frequency of backup. Personnel responsible for backup, a data backup schedule and a list of systems to be backed up are procedural details and will not be included in the data backup policy.

S3-257   Which of the following roles is **MOST** appropriately responsible for ensuring that security awareness and training material is effectively deployed to reach the intended audience?

    A. The human resources department
    B. The business manager
    C. The subject matter experts
    D. The information security department

D   The information security department oversees the information security program. This includes ensuring that training reaches the intended audience.

S3-258   Which of the following should be done **FIRST** when making a decision to allow access to the information processing facility (IPF) of an enterprise to a new external party?

A.   A contract language review
B.   A risk assessment
C.   The exposure factor
D.   Vendor due diligence

B   A risk assessment identifies the risks involved in allowing access to an external party and the required controls. The other choices could be part of the risk assessment.

S3-259   An enterprise has a network of suppliers that it allows to remotely access an important database that contains critical supply chain data. What is the **BEST** control to ensure that the individual supplier representatives who have access to the system do not improperly access or modify information within this system?

A.   User access rights
B.   Biometric access controls
C.   Password authentication
D.   Two-factor authentication

A   User access rights limit the access that users have to a network, file system or database once they have been authenticated. The remaining three responses are methods of user access control that manage user access to an overall system, not generally to a specific set of files or records.

S3-260   Which of the following is the **MOST** important consideration when developing a service level agreement (SLA) to mitigate the risk that outsourcing will result in a loss to the business?

A.   The nature of the indemnity clause
B.   Ensuring that the business objectives are defined and met
C.   Alignment of information system security objectives with enterprise goals
D.   Compliance with legal requirements

B   An indemnity clause is not the most important consideration and may not be part of the SLA. An SLA should be designed to deliver and protect the business needs. The security objective is what is being sought by implementing the control. While important, compliance with legal requirements is not generally a primary consideration for SLAs and, in many cases, is not a factor.

S3-261   When securing wireless access points, which of the following controls would **BEST** assure confidentiality?

A.   Implementing wireless intrusion prevention systems
B.   Not broadcasting the service set IDentifier (SSID)
C.   Implementing wired equivalent privacy (WEP) authentication
D.   Enforcing a virtual private network (VPN) over wireless

D   Enforcing a VPN over wireless is the best option to enforce strong authentication and encryption of the sessions. Implementing wireless intrusion prevention systems is a detective system and would not prevent wireless sniffing. Not broadcasting the SSID does not reduce the risk of wireless packets being captured. WEP authentication is known to be weak and does not protect individual confidentiality.

S3-262   A **PRIMARY** characteristic of a well-established information security culture is an alignment of:

    A. information security and business objectives.
    B. security controls with information technology.
    C. concurrent security strategies.
    D. values to protect corporate assets.

D   A culture is an environment within an enterprise where attitudes for its information security are shared across department lines and reflected in day-to-day activities. Choice A is a characteristic of an established information security program. Choices B and C are characteristics of an effective security architecture.

S3-263   Which of the following would **PRIMARILY** provide the potential for users to bypass a form-based authentication mechanism in an application with a back-end database?

    A. A weak password of six characters
    B. A structured query language (SQL) injection
    C. A session time-out of long duration
    D. Lack of an account lockout after multiple wrong attempts

B   Although SQL injection is well understood and preventable, it still is a significant security risk for many enterprises writing code. Using SQL injection, one can pass SQL statements in a manner that bypasses the logon page and allows access to the application. Weak passwords can make it easy to access the application by a weak authentication, but there is no bypass of authentication.

S3-264   Which of the following is the **MOST** important step when an employee is transferred to a different function?

    A. Reviewing and modifying access rights
    B. Assigning new security responsibilities
    C. Conducting specific training for the new role
    D. Knowledge of security weaknesses in last department

A   When an employee is transferred from one function to another, it is very important to review and update the logical access rights to ensure that any access no longer needed is removed and appropriate access for the new position is granted.

S3-265   Which of the following is the **BEST** way to erase confidential information stored on magnetic tapes?

    A. Performing a low-level format
    B. Rewriting with zeros
    C. Burning them
    D. Degaussing them

D   Degaussing the magnetic tapes would best dispose of confidential information since information is completely destroyed due to the magnetic effect of the degaussing process. Performing a low-level format and rewriting with zeros may still help, but some forensic tools can be used to retrieve information. Rewriting with zeros is dependent on the procedure used. Burning destroys the tapes and does not allow their reuse.

S3-266   The **MOST** common reason for an increasing number of emergency change requests is that:

A. the normal procedures are being bypassed.
B. there are zero-day defects.
C. there is an increase in help desk calls.
D. the IT team may be applying the changes without approval.

A   If there is an increasing number of emergency change requests, it means that people do not want to follow the standard process and thus the normal change control procedures are being bypassed.

S3-267   A contract has just been signed with a new vendor to manage IT support services. Which of the following tasks should the information security manager ensure is performed **NEXT**?

A. Establish vendor monitoring.
B. Define reporting relationships.
C. Create a service level agreement (SLA).
D. Have the vendor sign a nondisclosure agreement (NDA).

A   When a formal process has been followed, choices B, C and D are performed to define the parameters of the service relationship and provide the basis for establishing the contract. Once the contract is signed, the security manager should ensure that choice A, continuous vendor monitoring, is established and operational. This control will help identify and provide alerts on security events and minimize potential losses.

S3-268   Which of the following will be **MOST** important in calculating accurate return on investment (ROI) in information security?

A. Excluding qualitative risks for accuracy in calculated figures
B. Establishing processes to ensure cost reductions
C. Measuring monetary values in a consistent manner
D. Treating security investment as a profit center

C   There must be consistency in metrics in order to have accurate and consistent results. In assessing security risk, it is not a good idea to simply exclude qualitative risk because of the difficulties in measurement. If something is an important risk factor, an attempt should be made to quantify it even though it may not be highly accurate. ROI itself may not be primarily targeted for the assurance of cost reduction. Even when ROI is calculated, there is a chance that the security cost will increase if identified exposures are not immediately resolved. Treating a security investment as a profit center could be an important factor as an educational item for senior management. There is a fundamental requirement to run ROI-based security management, but it is not necessarily the key item in delivering positive results from ROI-based security management.

S3-269   Which of the following roles performs the day-to-day duties required to ensure the protection and integrity of data?

    A. Data owners
    B. Data users
    C. Steering committees
    D. Data custodians

D   Data custodians oversee the day-to-day duties required to ensure the protection and integrity of data. A custodian, such as IT systems personnel, may be responsible for performing regular backups, and testing the validity of backups and maintaining records in accordance with classification policies. In addition, a custodian may be the administrator for the enterprise's information classification scheme. Data owners decide the level of classification based upon business needs for the protection of data and periodically review the classification assignments and make changes as necessary. Information owners may be executives or managers responsible for the protection of data and may be held liable for negligence if there is a failure to protect the data. Data users follow procedures set out in the enterprise's security policy and adhere to privacy and security regulations that are often specific to sensitive application fields (e.g., health, finance, legal). Steering committees serve as an effective communication channel for management's aims and directions and provide an ongoing basis for ensuring alignment of the security program with enterprise objectives. Steering committees are also instrumental in achieving behavior change toward a culture that promotes good security practices and policy compliance.

S3-270   When outsourcing to an offshore provider, the **MOST** difficult element to determine during a security review will be:

    A. technical competency.
    B. incompatible culture.
    C. defense in depth.
    D. adequate policies.

B   Individuals in different cultures often have different perspectives on what information is considered sensitive or confidential and how the information should be handled. Those perspectives may not be consistent with the enterprise's requirements. Cultural norms are not usually an area of consideration in a security review or during an onsite inspection. Technical skills, controls and policies are the usual areas for review to ensure that they meet acceptable standards.

S3-271   Addressing production risks is **PRIMARILY** a function of:

    A. release management.
    B. incident management.
    C. change management.
    D. configuration management.

C   Change management is the overall process to assess and control risks introduced by changes. Release management is the specific process to manage risks of production system deployment. Incident management is not directly relevant to life-cycle stages. Configuration management is the specific process to manage risks associated with system configuration.

S3-272 Which of the following is the **BEST** approach to dealing with inadequate funding of the security program?

A. Eliminate low-priority security services.
B. Require management to accept the increased risk.
C. Prioritize risk mitigation and educate management.
D. Reduce monitoring and compliance enforcement activities.

C Allocating resources to the areas of highest risk and benefit and educating management on the potential consequences of underfunding is the best approach. Prioritizing security activities is always useful, but eliminating even low-priority security services should be a last resort. If budgets are seriously constrained, management is already addressing increases in other risks and is likely to be aware of the issue and a proactive approach to doing more with less will be well received. Reducing monitoring activities may unnecessarily increase risk when lower-cost options to perform those functions may be available.

S3-273 During an audit, an information security manager discovered that sales representatives are sending sensitive customer information through e-mail messages. Which of the following is the **BEST** course of action to address the issue?

A. Review the finding with the sales manager to evaluate the risk and impact.
B. Report the issue to senior management immediately.
C. Request that the sales representatives stop e-mailing sensitive information.
D. Provide security awareness training to the sales representatives.

A It is always good practice to engage the management of the business unit when addressing security threats and risks. The input from business unit management is critical in formulating the next step.

S3-274 Minimum requirements for database security settings are **BEST** defined through:

A. procedures.
B. guidelines.
C. baselines.
D. policies.

C Baselines set the minimum requirements. Procedures determine the steps, not the configuration requirements. Guidelines are not enforceable. Policies determine direction, but not detailed configurations.

S3-275 A set of metrics regarding the number of e-mail messages quarantined due to virus infection versus the number of infected e-mail messages that were not caught would be **MOST** useful to:

A. the security steering committee.
B. the board of directors.
C. IT managers.
D. the information security manager.

D Metrics support decisions. Knowing the number of e-mail messages blocked due to viruses would not on its own be an actionable piece of information for senior management (choices A and B) or for IT management (choice C). Information regarding the effectiveness of the current e-mail antivirus control is most useful to the information security manager and staff because they can use the information to initiate an investigation to determine why the control is not performing as expected and to determine whether there are other factors contributing to the failure of the control. When these determinations are made, the information security manager can use these metrics, along with data collected during the investigation, to support decisions to alter processes or add to (or change) the controls in place.

S3-276   To ensure that all employees follow procedures regarding the integrity and confidentiality of personal identifiable information (PII), a hospital required that policies and procedures be put in place for data access and that all data stored should be encrypted. This is an example of what type of controls?

      A.   Administrative and technical controls
      B.   Administrative and deterrent controls
      C.   Technical and physical controls
      D.   Administrative and corrective controls

A   Administrative and technical controls is the only correct answer.

S3-277   From an information security manager's perspective, which of the following is the **MOST** important element of a third-party contract to outsource a sensitive business process?

      A.   Security service level agreements (SLAs)
      B.   Background checks for key personnel
      C.   Specific system requirements
      D.   A right to audit

D   While SLAs, background checks and system requirements are important considerations, without the ability to audit the provider, it is very difficult to validate any of these other factors.

S3-278   An information security manager has implemented procedures for monitoring specific activities on the network. The system administrator has been trained to analyze the network events, take appropriate action and provide reports to the information security manager. What additional monitoring should be implemented to give a more accurate, risk-based view of network activity?

      A.   The system administrator should be monitored by a separate reviewer.
      B.   All activity on the network should be monitored.
      C.   No additional monitoring is needed in this situation.
      D.   Monitoring should be done only by the information security manager.

A   The system administrator needs to be monitored to ensure that the administrator is in compliance with the information security program. Normally, an administrator will have more rights on the network than an end user and, while an administrator can monitor others, administrators must be monitored as well. The primary objective is to ensure that risks are managed appropriately, balancing operational efficiency against adequate safety. To simply monitor all network activity is not a risk-based approach to protecting the enterprise. Choice C is not a true statement. The system administrator needs to be monitored for the specific activities. The information security manager needs to use the resources available within the enterprise to assist in monitoring compliance. Using expertise for monitoring is an efficient method and should be used when possible.

S3-279   Minimum standards for securing the technical infrastructure should be defined in a security:

      A.   strategy.
      B.   guidelines.
      C.   model.
      D.   architecture.

D   Minimum standards for securing the technical infrastructure should be defined in a security architecture document. This document defines how components are secured and the security services that should be in place. A strategy is a broad, high-level document. A guideline is advisory in nature, while a security model shows the relationships between components.

S3-280   It is **MOST** important that information security architecture be aligned with which of the following?

    A. Industry best practices
    B. Information technology plans
    C. Information security best practices
    D. Business objectives and goals

D   Information security architecture should always be properly aligned with business goals and objectives. Alignment with IT plans or industry and security best practices is secondary by comparison.

S3-281   Relationships among security technologies are **BEST** defined through which of the following?

    A. Security metrics
    B. Network topology
    C. Security architecture
    D. Process improvement models

C   Security architecture explains the use and relationships of security mechanisms. Security metrics measure improvement within the security practice but do not explain the use and relationships of security technologies. Process improvement models and network topology diagrams also do not describe the use and relationships of these technologies.

S3-282   Which of the following are likely to be updated **MOST** frequently?

    A. Procedures for hardening database servers
    B. Standards for password length and complexity
    C. Policies addressing information security governance
    D. Standards for document retention and destruction

A   Policies and standards should generally be more static and less subject to frequent change. Procedures on the other hand, especially with regard to the hardening of operating systems, will be subject to constant change; as operating systems change and evolve, the procedures for hardening will have to keep pace.

S3-283   Which of the following is the **MOST** important information to include in an information security standard?

    A. Creation date
    B. Author name
    C. Initial draft approval date
    D. Last review date

D   The last review date confirms the currency of the standard, affirming that management has reviewed the standard to assure that nothing in the environment has changed that would necessitate an update to the standard. The name of the author as well as the creation and draft dates are not that important.

S3-284   An information security manager at a global organization that is subject to regulation by multiple governmental jurisdictions with differing requirements should:

    A. bring all locations into conformity with the aggregate requirements of all governmental jurisdictions.
    B. establish baseline standards for all locations and add supplemental standards as required.
    C. bring all locations into conformity with a generally accepted set of industry best practices.
    D. establish a baseline standard incorporating those requirements that all jurisdictions have in common.

B   It is more efficient to establish a baseline standard and then develop additional standards for locations that must meet specific requirements. Seeking a lowest common denominator or just using industry best practices may cause certain locations to fail regulatory compliance. The opposite approach—forcing all locations to be in compliance with the regulations—places an undue burden on those locations.

S3-285   Reviewing which of the following would **BEST** ensure that security controls are effective?

    A. Risk assessment policies
    B. Return on security investment
    C. Security metrics
    D. User access rights

C   Reviewing security metrics provides senior management a snapshot view and trends of an organization's security posture. Choice A is incorrect because reviewing risk assessment policies would not ensure that the controls are actually working. Choice B is incorrect because reviewing returns on security investments provides business justifications in implementing controls, but does not measure effectiveness of the control itself. Choice D is incorrect because reviewing user access rights is a joint responsibility of the data custodian and the data owner, and does not measure control effectiveness.

S3-286   A good privacy statement should include:

    A. notification of liability on accuracy of information.
    B. notification that information will be encrypted.
    C. what the company will do with information it collects.
    D. a description of the information classification process.

C   Most privacy laws and regulations require disclosure on how information will be used. Choice A is incorrect because that information should be located in the web site's disclaimer. Choice B is incorrect because, although encryption may be applied, this is not generally disclosed. Choice D is incorrect because information classification would be contained in a separate policy.

S3-287   Which of the following would be **MOST** effective in successfully implementing restrictive password policies?

    A. Regular password audits
    B. Single sign-on system
    C. Security awareness program
    D. Penalties for noncompliance

C   To be successful in implementing restrictive password policies, it is necessary to obtain the buy-in of the end users. The best way to accomplish this is through a security awareness program. Regular password audits and penalties for noncompliance would not be as effective on their own; people would go around them unless forced by the system. Single sign-on is a technology solution that would enforce password complexity but would not promote user compliance. For the effort to be more effective, user buy-in is important.

S3-288   When an organization is setting up a relationship with a third-party IT service provider, which of the following is one of the **MOST** important topics to include in the contract from a security standpoint?

A. Compliance with international security standards.
B. Use of a two-factor authentication system.
C. Existence of an alternate hot site in case of business disruption.
D. Compliance with the organization's information security requirements.

D   From a security standpoint, compliance with the organization's information security requirements is one of the most important topics that should be included in the contract with third-party service provider. The scope of implemented controls in any ISO 27001-compliant organization depends on the security requirements established by each organization. Requiring compliance only with this security standard does not guarantee that a service provider complies with the organization's security requirements. The requirement to use a specific kind of control methodology is not usually stated in the contract with third-party service providers.

S3-289   When developing an information security program, what is the **MOST** useful source of information for determining available resources?

A. Proficiency test
B. Job descriptions
C. Organization chart
D. Skills inventory

D   A skills inventory would help identify the available resources, any gaps and the training requirements for developing resources. Proficiency testing is useful but only with regard to specific technical skills. Job descriptions would not be as useful since they may be out of date or not sufficiently detailed. An organization chart would not provide the details necessary to determine the resources required for this activity.

S3-290   Which of the following is an advantage of a centralized information security organizational structure?

A. It is easier to promote security awareness.
B. It is easier to manage and control.
C. It is more responsive to business unit needs.
D. It provides a faster turnaround for security requests.

B   It is easier to manage and control a centralized structure. Promoting security awareness is an advantage of decentralization. Decentralization allows you to use field security personnel as security missionaries or ambassadors to spread the security awareness message. Decentralized operations allow security administrators to be more responsive. Being close to the business allows decentralized security administrators to achieve a faster turnaround than that achieved in a centralized operation.

S3-291   The organization has decided to outsource the majority of the IT department with a vendor that is hosting servers in a foreign country. Of the following, which is the **MOST** critical security consideration?

     A.   Laws and regulations of the country of origin may not be enforceable in the foreign country.
     B.   A security breach notification might get delayed due to the time difference.
     C.   Additional network intrusion detection sensors should be installed, resulting in an additional cost.
     D.   The company could lose physical control over the server and be unable to monitor the physical security posture of the servers.

A   A company is held to the local laws and regulations of the country in which the company resides, even if the company decides to place servers with a vendor that hosts the servers in a foreign country. A potential violation of local laws applicable to the company might not be recognized or rectified (i.e., prosecuted) due to the lack of knowledge of the local laws that are applicable and the inability to enforce the laws. Choice B is not a problem. Time difference does not play a role in a 24/7 environment. Pagers, cellular phones, telephones, etc. are usually available to communicate notifications. Choice C is a manageable problem that requires additional funding, but can be addressed. Choice D is a problem that can be addressed. Most hosting providers have standardized the level of physical security that is in place. Regular physical audits or a SAS 70 report can address such concerns.

S3-292   Which of the following documents includes detailed requirements?

     A.   A policy
     B.   A guideline
     C.   A procedure
     D.   A standard

C   Procedures are detailed, step-by-step required actions. A policy is a statement of management intent. Guidelines are recommendations, but are not mandatory actions and are not typically detailed. A standard is a statement of a specific requirement, but it does not typically include details.

S3-293   One of the **MOST** likely benefits of decentralized security management is:

     A.   reduction of the total cost of ownership (TCO).
     B.   improved compliance with organizational policies and standards.
     C.   better alignment of security to business needs.
     D.   easier administration.

C   Better alignment of security to business needs is the only answer that fits since the other choices are benefits of centralized security management.

S3-294   Which would be one of the **BEST** metrics an information security manager can employ to effectively evaluate the results of a security program?

     A.   Number of controls implemented
     B.   Percent of control objectives accomplished
     C.   Percent of compliance with the security policy
     D.   Reduction in the number of reported security incidents

B   Control objectives are directly related to business objectives; therefore, they would be the best metrics. Number of controls implemented does not have a direct relationship with the results of a security program. Percentage of compliance with the security policy and reduction in the number of security incidents are not as broad as choice B.

S3-295   After obtaining commitment from senior management, which of the following should be completed **NEXT** when establishing an information security program?

    A. Define security metrics
    B. Conduct a risk assessment
    C. Perform a gap analysis
    D. Procure security tools

B   When establishing an information security program, conducting a risk assessment is key to identifying the needs of the organization and developing a security strategy. Defining security metrics, performing a gap analysis and procuring security tools are all subsequent considerations.

DOMAIN 4—INFORMATION SECURITY INCIDENT MANAGEMENT (18%)

S4-1   Which of the following should be determined **FIRST** when establishing a business continuity program?

A.   Cost to rebuild information processing facilities
B.   Incremental daily cost of the unavailability of systems
C.   Location and cost of offsite recovery facilities
D.   Composition and mission of individual recovery teams

B   Prior to creating a detailed business continuity plan, it is important to determine the incremental daily cost of losing different systems. This will allow recovery time objectives to be determined which, in turn, affects the location and cost of offsite recovery facilities, and the composition and mission of individual recovery teams. Determining the cost to rebuild information processing facilities would not be the first thing to determine.

S4-2   A company has a network of branch offices with local file/print and mail servers; each branch individually contracts a hot site. Which of the following would be the **GREATEST** weakness in recovery capability?

A.   Exclusive use of the hot site is limited to six weeks
B.   The hot site may have to be shared with other customers
C.   The time of declaration determines site access priority
D.   The provider services all major companies in the area

D   Sharing a hot site facility is sometimes necessary in the case of a major disaster. Also, first come, first served usually determines priority of access based on general industry practice. Access to a hot site is not indefinite; the recovery plan should address a long-term outage. In case of a disaster affecting a localized geographical area, the vendor's facility and capabilities could be insufficient for all of its clients, which will all be competing for the same resource. Preference will likely be given to the larger corporations, possibly delaying the recovery of a branch that will likely be smaller than other clients based locally.

S4-3   Which of the following actions should be taken when an online trading company discovers a network attack in progress?

A.   Shut off all network access points
B.   Dump all event logs to removable media
C.   Isolate the affected network segment
D.   Enable trace logging on all event

C   Isolating the affected network segment will mitigate the immediate threat while allowing unaffected portions of the business to continue processing. Shutting off all network access points would create a denial of service that could result in loss of revenue. Dumping event logs and enabling trace logging, while perhaps useful, would not mitigate the immediate threat posed by the network attack.

S4-4   The **BEST** method for detecting and monitoring a hacker's activities without exposing information assets
         to unnecessary risk is to utilize:

         A. firewalls.
         B. bastion hosts.
         C. decoy files.
         D. screened subnets.

C   Decoy files, often referred to as honeypots, are the best choice for diverting a hacker away from critical
         files and alerting security of the hacker's presence. Firewalls and bastion hosts attempt to keep the hacker
         out, while screened subnets or demilitarized zones (DMZs) provide a middle ground between the trusted
         internal network and the external untrusted Internet.

S4-5   The **FIRST** priority when responding to a major security incident is:

         A. documentation.
         B. monitoring.
         C. restoration.
         D. containment.

D   The first priority in responding to a security incident is to contain it to limit the impact. Documentation,
         monitoring and restoration are all important, but they should follow containment.

S4-6   Which of the following is the **MOST** important to ensure a successful recovery?

         A. Backup media is stored offsite
         B. Recovery location is secure and accessible
         C. More than one hot site is available
         D. Network alternate links are regularly tested

A   Unless backup media are available, all other preparations become meaningless. Recovery site location
         and security are important, but would not prevent recovery in a disaster situation. Having a secondary
         hot site is also important, but not as important as having backup media available. Similarly, alternate data
         communication lines should be tested regularly and successfully but, again, this is not as critical.

S4-7   Which of the following is the **MOST** important element to ensure the success of a disaster recovery test at
         a vendor-provided hot site?

         A. Tests are scheduled on weekends
         B. Network IP addresses are predefined
         C. Equipment at the hot site is identical
         D. Business management actively participates

D   Disaster recovery testing requires the allocation of sufficient resources to be successful. Without the
         support of management, these resources will not be available, and testing will suffer as a result. Testing
         on weekends can be advantageous but this is not the most important choice. As vendor-provided hot sites
         are in a state of constant change, it is not always possible to have network addresses defined in advance.
         Although it would be ideal to provide for identical equipment at the hot site, this is not always practical as
         multiple customers must be served and equipment specifications will therefore vary.

S4-8 At the conclusion of a disaster recovery test, which of the following should **ALWAYS** be performed prior to leaving the vendor's hot site facility?

A. Erase data and software from devices
B. Conduct a meeting to evaluate the test
C. Complete an assessment of the hot site provider
D. Evaluate the results from all test scripts

A For security and privacy reasons, all organizational data and software should be erased prior to departure. Evaluations can occur back at the office after everyone is rested, and the overall results can be discussed and compared objectively.

S4-9 An incident response policy must contain:

A. updated call trees.
B. escalation criteria.
C. press release templates.
D. critical backup files inventory.

B Escalation criteria, indicating the circumstances under which specific actions are to be undertaken, should be contained within an incident response policy. Telephone trees, press release templates and lists of critical backup files are too detailed to be included in a policy document.

S4-10 The **BEST** approach in managing a security incident involving a successful penetration should be to:

A. allow business processes to continue during the response.
B. allow the security team to assess the attack profile.
C. permit the incident to continue to trace the source.
D. examine the incident response process for deficiencies.

A Since information security objectives should always be linked to the objectives of the business, it is imperative that business processes be allowed to continue whenever possible. Only when there is no alternative should these processes be interrupted. Although it is important to allow the security team to assess the characteristics of an attack, this is subordinate to the needs of the business. Permitting an incident to continue may expose the organization to additional damage. Evaluating the incident management process for deficiencies is valuable but it, too, is subordinate to allowing business processes to continue.

S4-11 A post-incident review should be conducted by an incident management team to determine:

A. relevant electronic evidence.
B. lessons learned.
C. hacker's identity.
D. areas affected.

B Post-incident reviews are beneficial in determining ways to improve the response process through lessons learned from the attack. Evaluating the relevance of evidence, who launched the attack or what areas were affected are not the primary purposes for such a meeting because these should have been already established during the response to the incident.

S4-12   An organization with multiple data centers has designated one of its own facilities as the recovery site. The
         MOST important concern is the:

         A. communication line capacity between data centers.
         B. current processing capacity loads at data centers.
         C. differences in logical security at each center.
         D. synchronization of system software release versions.

B   If data centers are operating at or near capacity, it may prove difficult to recover critical operations at an
         alternate data center. Although line capacity is important from a mirroring perspective, this is secondary to
         having the necessary capacity to restore critical systems. By comparison, differences in logical and physical
         security and synchronization of system software releases are much easier issues to overcome and are,
         therefore, of less concern.

S4-13   Which of the following is **MOST** important in determining whether a disaster recovery test is successful?

         A. Only business data files from offsite storage are used
         B. IT staff fully recovers the processing infrastructure
         C. Critical business processes are duplicated
         D. All systems are restored within recovery time objectives (RTOs)

C   To ensure that a disaster recovery test is successful, it is most important to determine whether all critical
         business functions were successfully recovered and duplicated. Although ensuring that only materials taken
         from offsite storage are used in the test is important, this is not as critical in determining a test's success.
         While full recovery of the processing infrastructure is a key recovery milestone, it does not ensure the
         success of a test. Achieving the RTOs is another important milestone, but does not necessarily prove that
         the critical business functions can be conducted, due to interdependencies with other applications and key
         elements such as data, staff, manual processes, materials and accessories, etc.

S4-14   Which of the following is **MOST** important when deciding whether to build an alternate facility or
         subscribe to a third-party hot site?

         A. Cost to build a redundant processing facility and invocation
         B. Daily cost of losing critical systems and recovery time objectives (RTOs)
         C. Infrastructure complexity and system sensitivity
         D. Criticality results from the business impact analysis (BIA)

C   The complexity and business sensitivity of the processing infrastructure and operations largely determines
         the viability of such an option; the concern is whether the recovery site meets the operational and security
         needs of the organization. The cost to build a redundant facility is not relevant since only a fraction
         of the total processing capacity is considered critical at the time of the disaster and recurring contract
         costs would accrue over time. Invocation costs are not a factor because they will be the same regardless.
         The incremental daily cost of losing different systems and the recovery time objectives (RTOs) do not
         distinguish whether a commercial facility is chosen. Resulting criticality from the business impact analysis
         (BIA) will determine the scope and timeline of the recovery efforts, regardless of the recovery location.

S4-15   A new e-mail virus that uses an attachment disguised as a picture file is spreading rapidly over the Internet. Which of the following should be performed **FIRST** in response to this threat?

   A. Quarantine all picture files stored on file servers
   B. Block all e-mails containing picture file attachments
   C. Quarantine all mail servers connected to the Internet
   D. Block incoming Internet mail, but permit outgoing mail

B   Until signature files can be updated, incoming e-mail containing picture file attachments should be blocked. Quarantining picture files already stored on file servers is not effective since these files must be intercepted before they are opened. Quarantine of all mail servers or blocking all incoming mail is unnecessary overkill since only those e-mails containing attached picture files are in question.

S4-16   When a large organization discovers that it is the subject of a network probe, which of the following actions should be taken?

   A. Reboot the router connecting the DMZ to the firewall
   B. Power down all servers located on the DMZ segment
   C. Monitor the probe and isolate the affected segment
   D. Enable server trace logging on the affected segment

C   In the case of a probe, the situation should be monitored and the affected network segment isolated. Rebooting the router, powering down the demilitarized zone (DMZ) servers and enabling server trace routing are not warranted.

S4-17   Which of the following terms and conditions represent a significant deficiency if included in a commercial hot site contract?

   A. A hot site facility will be shared in multiple disaster declarations
   B. All equipment is provided "at time of disaster, not on floor"
   C. The facility is subject to a "first-come, first-served" policy
   D. Equipment may be substituted with equivalent model

B   Equipment provided "at time of disaster (ATOD), not on floor" means that the equipment is not available but will be acquired by the commercial hot site provider on a best effort basis. This leaves the customer at the mercy of the marketplace. If equipment is not immediately available, the recovery will be delayed. Many commercial providers do require sharing facilities in cases where there are multiple simultaneous declarations, and that priority may be established on a first-come, first-served basis. It is also common for the provider to substitute equivalent or better equipment, as they are frequently upgrading and changing equipment.

S4-18   Which of the following should be performed **FIRST** in the aftermath of a denial-of-service attack?

   A. Restore servers from backup media stored offsite
   B. Conduct an assessment to determine system status
   C. Perform an impact analysis of the outage
   D. Isolate the screened subnet

B   An assessment should be conducted to determine whether any permanent damage occurred and the overall system status. It is not necessary at this point to rebuild any servers. An impact analysis of the outage or isolating the demilitarized zone (DMZ) or screen subnet will not provide any immediate benefit.

S4-19   Which of the following is the **MOST** important element to ensure the successful recovery of a business during a disaster?

A. Detailed technical recovery plans are maintained offsite
B. Network redundancy is maintained through separate providers
C. Hot site equipment needs are recertified on a regular basis
D. Appropriate declaration criteria have been established

A   In a major disaster, staff can be injured or can be prevented from traveling to the hot site, so technical skills and business knowledge can be lost. It is therefore critical to maintain an updated copy of the detailed recovery plan at an offsite location. Continuity of the business requires adequate network redundancy, hot site infrastructure that is certified as compatible and clear criteria for declaring a disaster. Ideally, the business continuity program addresses all of these satisfactorily. However, in a disaster situation, where all these elements are present, but without the detailed technical plan, business recovery will be seriously impaired.

S4-20   The business continuity policy should contain which of the following?

A. Emergency call trees
B. Recovery criteria
C. Business impact assessment (BIA)
D. Critical backups inventory

B   Recovery criteria, indicating the circumstances under which specific actions are undertaken, should be contained within a business continuity policy. Telephone trees, business impact assessments (BIAs) and listings of critical backup files are too detailed to include in a policy document.

S4-21   The **PRIMARY** purpose of installing an intrusion detection system (IDS) is to identify:

A. weaknesses in network security.
B. patterns of suspicious access.
C. how an attack was launched on the network.
D. potential attacks on the internal network.

D   The most important function of an intrusion detection system (IDS) is to identify potential attacks on the network. Identifying how the attack was launched is secondary. It is not designed specifically to identify weaknesses in network security or to identify patterns of suspicious logon attempts.

S4-22   When an organization is using an automated tool to manage and house its business continuity plans, which of the following is the **PRIMARY** concern?

A. Ensuring accessibility should a disaster occur
B. Versioning control as plans are modified
C. Broken hyperlinks to resources stored elsewhere
D. Tracking changes in personnel and plan assets

A   If all of the plans exist only in electronic form, this presents a serious weakness if the electronic version is dependent on restoration of the intranet or other systems that are no longer available. Versioning control and tracking changes in personnel and plan assets is actually easier with an automated system. Broken hyperlinks are a concern, but less serious than plan accessibility.

S4-23   Which of the following is the **BEST** way to verify that all critical production servers are utilizing up-to-date virus signature files?

    A. Verify the date that signature files were last pushed out
    B. Use a recently identified benign virus to test if it is quarantined
    C. Research the most recent signature file and compare to the console
    D. Check a sample of servers that the signature files are current

D   The only accurate way to check the signature files is to look at a sample of servers. The fact that an update was pushed out to a server does not guarantee that it was properly loaded onto that server. Checking the vendor information to the management console would still not be indicative as to whether the file was properly loaded on the server. Personnel should never release a virus, no matter how benign.

S4-24   Which of the following actions should be taken when an information security manager discovers that a hacker is footprinting the network perimeter?

    A. Reboot the border router connected to the firewall
    B. Check IDS logs and monitor for any active attacks
    C. Update IDS software to the latest available version
    D. Enable server trace logging on the DMZ segment

B   Information security should check the intrusion detection system (IDS) logs and continue to monitor the situation. It would be inappropriate to take any action beyond that. In fact, updating the IDS could create a temporary exposure until the new version can be properly tuned. Rebooting the router and enabling server trace routing would not be warranted.

S4-25   Which of the following are the **MOST** important criteria when selecting virus protection software?

    A. Product market share and annualized cost
    B. Ability to interface with intrusion detection system (IDS) software and firewalls
    C. Alert notifications and impact assessments for new viruses
    D. Ease of maintenance and frequency of updates

D   For the software to be effective, it must be easy to maintain and keep current. Market share and annualized cost, links to the intrusion detection system (IDS) and automatic notifications are all secondary in nature.

S4-26   Which of the following is the **MOST** serious exposure of automatically updating virus signature files on every desktop each Friday at 11:00 p.m. (23.00 hrs.)?

    A. Most new viruses' signatures are identified over weekends
    B. Technical personnel are not available to support the operation
    C. Systems are vulnerable to new viruses during the intervening week
    D. The update's success or failure is not known until Monday

C   Updating virus signature files on a weekly basis carries the risk that the systems will be vulnerable to viruses released during the week; far more frequent updating is essential. All other issues are secondary to this very serious exposure.

S4-27   When performing a business impact analysis (BIA), which of the following should calculate the recovery time and cost estimates?

A. Business continuity coordinator
B. Information security manager
C. Business process owners
D. IT management

C   Business process owners are in the best position to understand the true impact on the business that a system outage would create. The business continuity coordinator, IT management and even the information security manager will not be able to provide that level of detailed knowledge.

S4-28   Which of the following is **MOST** closely associated with a business continuity program?

A. Confirming that detailed technical recovery plans exist
B. Periodically testing network redundancy
C. Updating the hot site equipment configuration every quarter
D. Developing recovery time objectives (RTOs) for critical functions

D   Technical recovery plans, network redundancy and equipment needs are all associated with infrastructure disaster recovery. Only recovery time objectives (RTOs) directly relate to business continuity.

S4-29   Which of the following application systems should have the shortest recovery time objective (RTO)?

A. Contractor payroll
B. Change management
C. E-commerce web site
D. Fixed asset system

C   In most businesses where an e-commerce site is in place, it would need to be restored in a matter of hours, if not minutes. Contractor payroll, change management and fixed assets would not require as rapid a recovery time.

S4-30   A computer incident response team (CIRT) manual should **PRIMARILY** contain which of the following documents?

A. Risk assessment results
B. Severity criteria
C. Emergency call tree directory
D. Table of critical backup files

B   Quickly ranking the severity criteria of an incident is a key element of incident response. The other choices refer to documents that would not likely be included in a computer incident response team (CIRT) manual.

S4-31   Which of the following would represent a violation of the chain of custody when a backup tape has been identified as evidence in a fraud investigation? The tape was:

A. removed into the custody of law enforcement investigators.
B. kept in the tape library pending further analysis.
C. sealed in a signed envelope and locked in a safe under dual control.
D. handed over to authorized independent investigators.

B   Since a number of individuals would have access to the tape library, and could have accessed and tampered with the tape, the chain of custody could not be verified. All other choices provide clear indication of who was in custody of the tape at all times.

S4-32   When properly tested, which of the following would **MOST** effectively support an information security manager in handling a security breach?

A. Business continuity plan
B. Disaster recovery plan
C. Incident response plan
D. Vulnerability management plan

C   An incident response plan documents the step-by-step process to follow, as well as the related roles and responsibilities pertaining to all parties involved in responding to an information security breach. A business continuity plan or disaster recovery plan would be triggered during the execution of the incident response plan in the case of a breach impacting the business continuity. A vulnerability management plan is a procedure to address technical vulnerabilities and mitigate the risk through configuration changes (patch management).

S4-33   Isolation and containment measures for a compromised computer have been taken and information security management is now investigating. What is the **MOST** appropriate next step?

A. Run a forensics tool on the machine to gather evidence
B. Reboot the machine to break remote connections
C. Make a copy of the whole system's memory
D. Document current connections and open Transmission Control Protocol/User Datagram Protocol (TCP/UDP) ports

C   When investigating a security breach, it is important to preserve all traces of evidence left by the invader. For this reason, it is imperative to preserve the memory contents of the machine in order to analyze them later. The correct answer is choice C because a copy of the whole system's memory is obtained for future analysis by running the appropriate tools. This is also important from a legal perspective since an attorney may suggest that the system was changed during the conduct of the investigation. Running a computer forensics tool in the compromised machine will cause the creation of at least one process that may overwrite evidence. Rebooting the machine will delete the contents of the memory, erasing potential evidence. Collecting information about current connections and open Transmission Control Protocol/User Datagram Protocol (TCP/UDP) ports is correct, but doing so by using tools may also erase memory contents.

S4-34   Why is "slack space" of value to an information security manager as part of an incident investigation?

    A. Hidden data may be stored there
    B. The slack space contains login information
    C. Slack space is encrypted
    D. It provides flexible space for the investigation

A   "Slack space" is the unused space between where the file data end and the end of the cluster the data occupy. Login information is not typically stored in the slack space. Encryption for the slack space is no different from the rest of the file system. The slack space is not a viable means of storage during an investigation.

S4-35   What is the **PRIMARY** objective of a post-event review in incident response?

    A. Adjust budget provisioning
    B. Preserve forensic data
    C. Improve the response process
    D. Ensure the incident is fully documented

C   The primary objective is to find any weakness in the current process and improve it. The other choices are all secondary.

S4-36   Detailed business continuity plans should be based **PRIMARILY** on:

    A. consideration of different alternatives.
    B. the solution that is least expensive.
    C. strategies that cover all applications.
    D. strategies validated by senior management.

D   A recovery strategy identifies the best way to recover a system in case of disaster and provides guidance based on detailed recovery procedures that can be developed. Different strategies should be developed and all alternatives presented to senior management. Senior management should select the most appropriate strategy from the alternatives provided. The selected strategy should be used for further development of the detailed business continuity plan. The selection of strategy depends on criticality of the business process and applications supporting the processes. It need not necessarily cover all applications. All recovery strategies have associated costs, which include costs of preparing for disruptions and putting them to use in the event of a disruption. The latter can be insured against, but not the former. The best recovery option need not be the least expensive.

S4-37   A web server in a financial institution that has been compromised using a super-user account has been isolated, and proper forensic processes have been followed. The next step should be to:

    A. rebuild the server from the last verified backup.
    B. place the web server in quarantine.
    C. shut down the server in an organized manner.
    D. rebuild the server with original media and relevant patches.

D   The original media should be used since one can never be sure of all the changes a super-user may have made nor the timelines in which these changes were made. Rebuilding from the last known verified backup is incorrect since the verified backup may have been compromised by the super-user at a different time. Placing the web server in quarantine should have already occurred in the forensic process. Shut down in an organized manner is out of sequence and no longer a problem. The forensic process is already finished and evidence has already been acquired.

S4-38   Evidence from a compromised server has to be acquired for a forensic investigation. What would be the **BEST** source?

A.   A bit-level copy of all hard drive data
B.   The last verified backup stored offsite
C.   Data from volatile memory
D.   Backup servers

A   The bit-level copy image file ensures forensic quality evidence that is admissible in a court of law. Choices B and D may not provide forensic quality data for investigative work, while choice C alone may not provide enough evidence.

S4-39   In the course of responding to an information security incident, the **BEST** way to treat evidence for possible legal action is defined by:

A.   international standards.
B.   local regulations.
C.   generally accepted best practices.
D.   organizational security policies.

B   Legal follow-up will most likely be performed locally where the incident took place; therefore, it is critical that the procedure of treating evidence is in compliance with local regulations. In certain countries, there are strict regulations on what information can be collected. When evidence collected is not in compliance with local regulations, it may not be admissible in court. There are no common regulations to treat computer evidence that are accepted internationally. Generally accepted best practices such as a common chain-of-custody concept may have different implementation in different countries, and thus may not be a good assurance that evidence will be admissible. Local regulations always take precedence over organizational security policies.

S4-40   Emergency actions are taken at the early stage of a disaster with the purpose of preventing injuries or loss of life and:

A.   determining the extent of property damage.
B.   preserving environmental conditions.
C.   ensuring orderly plan activation.
D.   reducing the extent of operational damage.

D   During an incident, emergency actions should minimize or eliminate casualties and damage to the business operation, thus reducing business interruptions. Determining the extent of property damage is not the consideration; emergency actions should minimize, not determine, the extent of the damage. Protecting/ preserving environmental conditions may not be relevant. Ensuring orderly plan activation is important but not as critical as reducing damage to the operation.

S4-41   What is the **FIRST** action an information security manager should take when a company laptop is reported stolen?
A.   Evaluate the impact of the information loss
B.   Update the corporate laptop inventory
C.   Ensure compliance with reporting procedures
D.   Disable the user account immediately

C   The first step is to ensure proper reporting procedures are followed. The other steps would follow.

S4-42   Which of the following actions should take place immediately after a security breach is reported to an
         information security manager?

         A. Confirm the incident
         B. Determine impact
         C. Notify affected stakeholders
         D. Isolate the incident

A   Before performing analysis of impact, resolution, notification or isolation of an incident, it must be
         validated as a real security incident.

S4-43   When designing the technical solution for a disaster recovery site, the **PRIMARY** factor that should be
         taken into consideration is the:

         A. services delivery objective.
         B. recovery time objective (RTO).
         C. recovery window.
         D. maximum tolerable outage (MTO).

C   The length of the recovery window is defined by business management and determines the acceptable time
         frame between a disaster and the restoration of critical services/applications. The technical implementation
         of the disaster recovery (DR) site will be based on this constraint, especially the choice between a hot,
         warm or cold site. The service delivery objective is supported during the alternate process mode until the
         normal situation is restored, which is directly related to business needs. The recovery time objective (RTO)
         is commonly agreed to be the time frame between a disaster and the return to normal operations. It is then
         longer than the interruption window and is very difficult to estimate in advance. The time frame between
         the reduced operation mode at the end of the interruption window and the return to normal operations
         depends on the magnitude of the disaster. Technical disaster recovery solutions alone will not be used for
         returning to normal operations. Maximum tolerable outage (MTO) is the maximum time acceptable by a
         company operating in reduced mode before experiencing losses. Theoretically, recovery time objectives
         (RTOs) equal the interruption window plus the maximum tolerable outage. This will not be the primary
         factor for the choice of the technical disaster recovery solution.

S4-44   In designing a backup strategy that will be consistent with a disaster recovery strategy, the **PRIMARY**
         factor to be taken into account will be the:

         A. volume of sensitive data.
         B. recovery point objective (RPO).
         C. recovery time objective (RTO).
         D. interruption window.

B   The recovery point objective (RPO) defines the maximum loss of data (in terms of time) acceptable by
         the business (i.e., age of data to be restored). It will directly determine the basic elements of the backup
         strategy—frequency of the backups and what kind of backup is the most appropriate (disk-to-disk, on tape,
         mirroring). The volume of data will be used to determine the capacity of the backup solution. The recovery
         time objective (RTO)—the time between disaster and return to normal operation—will not have any impact
         on the backup strategy. The availability to restore backups in a time frame consistent with the interruption
         window will have to be checked and will influence the strategy (e.g., full backup vs. incremental), but this
         will not be the primary factor.

S4-45   An intrusion detection system (IDS) should:

    A. run continuously
    B. ignore anomalies
    C. require a stable, rarely changed environment
    D. be located on the network

A   If an intrusion detection system (IDS) does not run continuously the business remains vulnerable. An IDS should detect, not ignore anomalies. An IDS should be flexible enough to cope with a changing environment. Both host and network based IDS are recommended for adequate detection.

S4-46   The **PRIORITY** action to be taken when a server is infected with a virus is to:

    A. isolate the infected server(s) from the network.
    B. identify all potential damage caused by the infection.
    C. ensure that the virus database files are current.
    D. establish security weaknesses in the firewall.

A   The priority in this event is to minimize the effect of the virus infection and to prevent it from spreading by removing the infected server(s) from the network. After the network is secured from further infection, the damage assessment can be performed, the virus database updated and any weaknesses sought.

S4-47   Which of the following provides the **BEST** confirmation that the business continuity/disaster recovery plan objectives have been achieved?

    A. The recovery time objective (RTO) was not exceeded during testing
    B. Objective testing of the business continuity/disaster recovery plan has been carried out consistently
    C. The recovery point objective (RPO) was proved inadequate by disaster recovery plan testing
    D. Information assets have been valued and assigned to owners per the business continuity plan/disaster recovery plan

A   Consistent achievement of recovery time objective (RTO) objectives during testing provides the most objective evidence that business continuity/disaster recovery plan objectives have been achieved. The successful testing of the business continuity/disaster recovery plan within the stated RTO objectives is the most indicative evidence that the business needs are being met. Objective testing of the business continuity/ disaster recovery plan will not serve as a basis for evaluating the alignment of the risk management process in business continuity/disaster recovery planning.. Mere valuation and assignment of information assets to owners (per the business continuity/disaster recovery plan) will not serve as a basis for evaluating the alignment of the risk management process in business continuity/disaster recovery planning.

S4-48   Which of the following situations would be the **MOST** concern to a security manager?

    A. Audit logs are not enabled on a production server
    B. The logon ID for a terminated systems analyst still exists on the system
    C. The help desk has received numerous results of users receiving phishing e-mails
    D. A Trojan was found to be installed on a system administrator's laptop

D   The discovery of a Trojan installed on a system's administrator's laptop is highly significant since this may mean that privileged user accounts and passwords may have been compromised. The other choices, although important, do not pose as immediate or as critical a threat.

S4-49   A customer credit card database has been reported as being breached by hackers. The **FIRST** step in dealing with this attack should be to:

    A. confirm the incident.
    B. notify senior management.
    C. start containment.
    D. notify law enforcement.

A   Asserting that the condition is a true security incident is the necessary first step in determining the correct response. The containment stage would follow. Notifying senior management and law enforcement could be part of the incident response process that takes place after confirming an incident.

S4-50   A root kit was used to capture detailed accounts receivable information. To ensure admissibility of evidence from a legal standpoint, once the incident was identified and the server isolated, the next step should be to:

    A. document how the attack occurred.
    B. notify law enforcement.
    C. take an image copy of the media.
    D. close the accounts receivable system.

C   Taking an image copy of the media is a recommended practice to ensure legal admissibility. All of the other choices are subsequent and may be supplementary.

S4-51   When collecting evidence for forensic analysis, it is important to:

    A. ensure the assignment of qualified personnel.
    B. request the IT department do an image copy.
    C. disconnect from the network and isolate the affected devices.
    D. ensure law enforcement personnel are present before the forensic analysis commences.

A   Without the initial assignment of forensic expertise, the required levels of evidence may not be preserved. In choice B, the IT department is unlikely to have that level of expertise and should, thus, be prevented from taking action. Choice C may be a subsequent necessity that comes after choice A. Choice D, notifying law enforcement, will likely occur after the forensic analysis has been completed.

S4-52   What is the **BEST** method for mitigating against network denial of service (DoS) attacks?

    A. Ensure all servers are up-to-date on OS patches
    B. Employ packet filtering to drop suspect packets
    C. Implement network address translation to make internal addresses nonroutable
    D. Implement load balancing for Internet facing devices

B   Packet filtering techniques are the only ones which reduce network congestion caused by a network denial of service (DoS) attack. Patching servers, in general, will not affect network traffic. Implementing network address translation and load balancing would not be as effective in mitigating most network DoS attacks.

S4-53   To justify the establishment of an incident management team, an information security manager would find which of the following to be the **MOST** effective?

     A.   Assessment of business impact of past incidents
     B.   Need of an independent review of incident causes
     C.   Need for constant improvement on the security level
     D.   Possible business benefits from incident impact reduction

D   Business benefits from incident impact reduction would be the most important goal for establishing an incident management team. The assessment of business impact of past incidents would need to be completed to articulate the benefits. Having an independent review benefits the incident management process. The need for constant improvement on the security level is a benefit to the organization.

S4-54   A database was compromised by guessing the password for a shared administrative account and confidential customer information was stolen. The information security manager was able to detect this breach by analyzing which of the following?

     A.   Invalid logon attempts
     B.   Write access violations
     C.   Concurrent logons
     D.   Firewall logs

A   Since the password for the shared administrative account was obtained through guessing, it is probable that there were multiple unsuccessful logon attempts before the correct password was deduced. Searching the logs for invalid logon attempts could, therefore, lead to the discovery of this unauthorized activity. Because the account is shared, reviewing the logs for concurrent logons would not reveal unauthorized activity since concurrent usage is common in this situation. Write access violations would not necessarily be observed since the information was merely copied and not altered. Firewall logs would not necessarily contain information regarding logon attempts.

S4-55   Which of the following is an example of a corrective control?

     A.   Diverting incoming traffic as a response to a denial of service (DoS) attack
     B.   Filtering network traffic
     C.   Examining inbound network traffic for viruses
     D.   Logging inbound network traffic

A   Diverting incoming traffic corrects the situation and, therefore, is a corrective control. Choice B is a preventive control. Choices C and D are detective controls.

S4-56   To determine how a security breach occurred on the corporate network, a security manager looks at the logs of various devices. Which of the following **BEST** facilitates the correlation and review of these logs?

     A.   Database server
     B.   Domain name server (DNS)
     C.   Time server
     D.   Proxy server

C   To accurately reconstruct the course of events, a time reference is needed and that is provided by the time server. The other choices would not assist in the correlation and review of these logs.

S4-57   An organization has been experiencing a number of network-based security attacks that all appear to originate internally. The **BEST** course of action is to:

A.   require the use of strong passwords.
B.   assign static IP addresses.
C.   implement centralized logging software.
D.   install an intrusion detection system (IDS).

D   Installing an intrusion detection system (IDS) will allow the information security manager to better pinpoint the source of the attack so that countermeasures may then be taken. An IDS is not limited to detection of attacks originating externally. Proper placement of agents on the internal network can be effectively used to detect an internally based attack. Requiring the use of strong passwords will not be sufficiently effective against a network-based attack. Assigning IP addresses would not be effective since these can be spoofed. Implementing centralized logging software will not necessarily provide information on the source of the attack.

S4-58   A serious vulnerability is reported in the firewall software used by an organization. Which of the following should be the immediate action of the information security manager?

A.   Ensure that all OS patches are up-to-date
B.   Block inbound traffic until a suitable solution is found
C   Obtain guidance from the firewall manufacturer
D.   Commission a penetration test

C   The best source of information is the firewall manufacturer since the manufacturer may have a patch to fix the vulnerability or a workaround solution. Ensuring that all OS patches are up-to-date is a best practice, in general, but will not necessarily address the reported vulnerability. Blocking inbound traffic may not be practical or effective from a business perspective. Commissioning a penetration test will take too much time and will not necessarily provide a solution for corrective actions.

S4-59   An organization keeps backup tapes of its servers at a warm site. To ensure that the tapes are properly maintained and usable during a system crash, the **MOST** appropriate measure the organization should perform is to:

A.   use the test equipment in the warm site facility to read the tapes.
B.   periodically retrieve the tapes from the warm site and test them.
C.   have duplicate equipment available at the warm site.
D.   inspect the facility and inventory the tapes on a quarterly basis.

B   A warm site is not fully equipped with the company's main systems; therefore, the tapes should be periodically tested using the company's production systems. Inspecting the facility and checking the tape inventory does not guarantee that the tapes are usable.

S4-60   Which of the following processes is critical for deciding prioritization of actions in a business continuity plan?

    A. Business impact analysis (BIA)
    B. Risk assessment
    C. Vulnerability assessment
    D. Business process mapping

A   A business impact analysis (BIA) provides results, such as impact from a security incident and required response times. The BIA is the most critical process for deciding which part of the information system/business process should be given prioritization in case of a security incident. Risk assessment is a very important process for the creation of a business continuity plan. Risk assessment provides information on the likelihood of occurrence of security incidence and assists in the selection of countermeasures, but not in the prioritization. As in choice B, a vulnerability assessment provides information regarding the security weaknesses of the system, supporting the risk analysis process. Business process mapping facilitates the creation of the plan by providing mapping guidance on actions after the decision on critical business processes has been made–translating business prioritization to IT prioritization. Business process mapping does not help in making a decision, but in implementing a decision.

S4-61   In addition to backup data, which of the following is the **MOST** important to store offsite in the event of a disaster?

    A. Copies of critical contracts and service level agreements (SLAs)
    B. Copies of the business continuity plan
    C. Key software escrow agreements for the purchased systems
    D. List of emergency numbers of service providers

B   Without a copy of the business continuity plan, recovery efforts would be severely hampered or may not be effective. All other choices would not be as immediately critical as the business continuity plan itself. The business continuity plan would contain a list of the emergency numbers of service providers.

S4-62   Which of the following is the **MOST** important consideration for an organization interacting with the media during a disaster?

    A. Communicating specially drafted messages by an authorized person
    B. Refusing to comment until recovery
    C. Referring the media to the authorities
    D. Reporting the losses and recovery strategy to the media

A   Proper messages need to be sent quickly through a specific identified person so that there are no rumors or statements made that may damage reputation. Choices B, C and D are not recommended until the message to be communicated is made clear and the spokesperson has already spoken to the media.

S4-63   During the security review of organizational servers it was found that a file server containing confidential human resources (HR) data was accessible to all user IDs. As a **FIRST** step, the security manager should:

A. copy sample files as evidence.
B. remove access privileges to the folder containing the data.
C. report this situation to the data owner.
D. train the HR team on properly controlling file permissions.

C   The data owner should be notified prior to any action being taken. Copying sample files as evidence is not advisable since it breaches confidentiality requirements on the file. Removing access privileges to the folder containing the data should be done by the data owner or by the security manager in consultation with the data owner—frequently the security manager would not have this right anyway; regardless, this would be done only after formally reporting the incident. Training the human resources (HR) team on properly controlling file permissions is the method to prevent such incidents in the future, but should take place once the incident reporting and investigation activities are completed.

S4-64   If an organization considers taking legal action on a security incident, the information security manager should focus **PRIMARILY** on:

A. obtaining evidence as soon as possible.
B. preserving the integrity of the evidence.
C. disconnecting all IT equipment involved.
D. reconstructing the sequence of events.

B   The integrity of evidence should be kept, following the appropriate forensic techniques to obtain the evidence and a chain of custody procedure to maintain the evidence (in order to be accepted in a court of law). All other options are part of the investigative procedure, but they are not as important as preserving the integrity of the evidence.

S4-65   Which of the following has the highest priority when defining an emergency response plan?

A. Critical data
B. Critical infrastructure
C. Safety of personnel
D. Vital records

C   The safety of an organization's employees should be the most important consideration given human safety laws. Human safety is considered first in any process or management practice. All of the other choices are secondary.

S4-66   The **PRIMARY** purpose of involving third-party teams for carrying out postevent reviews of information security incidents is to:

A. enable independent and objective review of the root cause of the incidents.
B. obtain support for enhancing the expertise of the third-party teams.
C. identify lessons learned for further improving the information security management process.
D. obtain better buy-in for the information security program.

A   It is always desirable to avoid the conflict of interest involved in having the information security team carry out the postevent review. Obtaining support for enhancing the expertise of the third-party teams is one of the advantages, but is not the primary driver. Identifying lessons learned for further improving the information security management process is the general purpose of carrying out the postevent review. Obtaining better buy-in for the information security program is not a valid reason for involving third-party teams.

S4-67   The **MOST** important objective of a postincident review is to:

A. capture lessons learned to improve the process.
B. develop a process for continuous improvement.
C. develop a business case for the security program budget.
D. identify new incident management tools.

A   The main purpose of a postincident review is to identify areas of improvement in the process. Developing a process for continuous improvement is not true in every case. Developing a business case for the security program budget and identifying new incident management tools may come from the analysis of the incident, but are not the key objectives.

S4-68   Which of the following is the **MOST** critical consideration when collecting and preserving admissible evidence during an incident response?

A. Unplugging the systems
B. Chain of custody
C. Separation of duties
D. Clock synchronization

B   Admissible evidence must be collected and preserved by "chain of custody." Unplugging the systems can cause potential loss of information critical to the investigation. Separation of duties is not necessary in evidence collection and preservation since the entire process can be done by a single person. Clock synchronization is not as important for the collection and preservation of admissible evidence.

S4-69   In a forensic investigation, which of the following would be the **MOST** important factor?

A. Operation of a robust incident management process
B. Identification of areas of responsibility
C. Involvement of law enforcement
D. Expertise of resources

D   The most important factor in a forensic investigation is the expertise of the resources participating in the project due to the inherent complexity. Operation of a robust incident management process and the identification of areas of responsibility should occur prior to an investigation. Involvement of law enforcement is dependent upon the nature of the investigation.

S4-70   When a major vulnerability in the security of a critical web server is discovered, immediate notification should be made to the:

    A.   system owner to take corrective action.
    B.   incident response team to investigate.
    C.   data owners to mitigate damage.
    D.   development team to remediate.

A   In order to correct the vulnerabilities, the system owner needs to be notified quickly before an incident can take place. Choice B is not correct because the incident has not taken place and notification could delay implementation of the fix. Data owners would be notified only if the vulnerability could have compromised data. The development team may be called upon by the system owner to resolve the vulnerability.

S4-71   Three employees reported the theft or loss of their laptops while on business trips. The **FIRST** course of action for the security manager is to:

    A.   assess the impact of the loss and determine mitigating steps.
    B.   communicate the best practices in protecting laptops to all laptop users.
    C.   instruct the erring employees to pay a penalty for the lost laptops.
    D.   recommend that management report the incident to the police and file for insurance.

A   The first step when addressing theft or loss is to determine what was actually lost and the appropriate response. Choice B may occur after the impact is assessed. Choices C and D depend upon company policy.

S4-72   Which of the following is the **BEST** mechanism to determine the effectiveness of the incident response process?

    A.   Incident response metrics
    B.   Periodic auditing of the incident response process
    C.   Action recording and review
    D.   Postincident review

D   Postevent reviews are designed to identify gaps and shortcomings in the actual incident response process so that these gaps may be improved over time. The other choices will not provide the same level of feedback in improving the process.

S4-73   The **FIRST** step in an incident response plan is to:

    A.   notify the appropriate individuals.
    B.   contain the effects of the incident to limit damage.
    C.   develop response strategies for systematic attacks.
    D.   validate the incident.

D   Appropriate people need to be notified; however, one must first validate the incident. Containing the effects of the incident would be completed after validating the incident. Developing response strategies for systematic attacks should have already been developed prior to the occurrence of an incident.

S4-74   An organization has verified that its customer information was recently exposed. Which of the following is the **FIRST** step a security manager should take in this situation?

A.   Inform senior management.
B.   Determine the extent of the compromise.
C.   Report the incident to the authorities.
D.   Communicate with the affected customers.

B   Before reporting to senior management, affected customers or the authorities, the extent of the exposure needs to be assessed.

S4-75   A possible breach of an organization's IT system is reported by the project manager. What is the **FIRST** thing the incident response manager should do?

A.   Run a port scan on the system
B.   Disable the logon ID
C.   Investigate the system logs
D.   Validate the incident

D   When investigating a possible incident, it should first be validated. Running a port scan on the system, disabling the logon IDs and investigating the system logs may be required based on preliminary forensic investigation, but doing so as a first step may destroy the evidence.

S4-76   The **PRIMARY** consideration when defining recovery time objectives (RTOs) for information assets is:

A.   regulatory requirements.
B.   business requirements.
C.   financial value.
D.   IT resource availability.

B   The criticality to business should always drive the decision. Regulatory requirements could be more flexible than business needs. The financial value of an asset could not correspond to its business value. While a consideration, IT resource availability is not a primary factor.

S4-77   What task should be performed once a security incident has been verified?

A.   Identify the incident.
B.   Contain the incident.
C.   Determine the root cause of the incident.
D.   Perform a vulnerability assessment.

B   Identifying the incident means verifying whether an incident has occurred and finding out more details about the incident. Once an incident has been confirmed (identified), the incident management team should limit further exposure. Determining the root cause takes place after the incident has been contained. Performing a vulnerability assessment takes place after the root cause of an incident has been determined, in order to find new vulnerabilities.

S4-78 An information security manager believes that a network file server was compromised by a hacker. Which of the following should be the **FIRST** action taken?

    A. Ensure that critical data on the server are backed up.
    B. Shut down the compromised server.
    C. Initiate the incident response process.
    D. Shut down the network.

C The incident response process will determine the appropriate course of action. If the data have been corrupted by a hacker, the backup may also be corrupted. Shutting down the server is likely to destroy any forensic evidence that may exist and may be required by the investigation. Shutting down the network is a drastic action, especially if the hacker is no longer active on the network.

S4-79 An unauthorized user gained access to a merchant's database server and customer credit card information. Which of the following would be the **FIRST** step to preserve and protect the evidence of unauthorized intrusion activities?

    A. Shut down and power off the server.
    B. Duplicate the hard disk of the server immediately.
    C. Isolate the server from the network.
    D. Copy the database log file to a protected server.

C Isolating the server will prevent further intrusions and protect evidence of intrusion activities left in memory and on the hard drive. Some intrusion activities left in virtual memory may be lost if the system is shut down. Duplicating the hard disk will only preserve the evidence on the hard disk, not the evidence in virtual memory, and will not prevent further unauthorized access attempts. Copying the database log file to a protected server will not provide sufficient evidence should the organization choose to pursue legal recourse.

S4-80 Which of the following would be a **MAJOR** consideration for an organization defining its business continuity plan (BCP) or disaster recovery program (DRP)?

    A. Setting up a backup site
    B. Maintaining redundant systems
    C. Aligning with recovery time objectives (RTOs)
    D. Data backup frequency

C BCP/DRP should align with business RTOs. The RTO represents the amount of time allowed for the recovery of a business function or resource after a disaster occurs. The RTO must be taken into consideration when prioritizing systems for recovery efforts to ensure that those systems that the business requires first are the ones that are recovered first.

S4-81 Which of the following would be **MOST** appropriate for collecting and preserving evidence?

    A. Encrypted hard drives
    B. Generic audit software
    C. Proven forensic processes
    D. Log correlation software

C When collecting evidence about a security incident, it is very important to follow appropriate forensic procedures to handle electronic evidence by a method approved by local jurisdictions. All other options will help when collecting or preserving data about the incident; however these data might not be accepted as evidence in a court of law if they are not collected by a method approved by local jurisdictions.

S4-82   Which of the following is the **MOST** important aspect of forensic investigations that will potentially involve legal action?

A. The independence of the investigator
B. Timely intervention
C. Identifying the perpetrator
D. Chain of custody

D   Establishing the chain of custody is one of the most important steps in conducting forensic investigations since it preserves the evidence in a manner that is admissible in court. The independence of the investigator may be important, but is not the most important aspect. Timely intervention is important for containing incidents, but not as important for forensic investigation. Identifying the perpetrator is important, but maintaining the chain of custody is more important in order to have the perpetrator convicted in court.

S4-83   In the course of examining a computer system for forensic evidence, data on the suspect media were inadvertently altered. Which of the following should have been the **FIRST** course of action in the investigative process?

A. Perform a backup of the suspect media to new media.
B. Create a bit-by-bit image of the original media source onto new media.
C. Make a copy of all files that are relevant to the investigation.
D. Run an error-checking program on all logical drives to ensure that there are no disk errors.

B   The original hard drive or suspect media should never be used as the source for analysis. The source or original media should be physically secured and only used as the master to create a bit-by-bit image. The original should be stored using the appropriate procedures, depending on location. The image created for forensic analysis should be used. A backup does not preserve 100 percent of the data, such as erased or deleted files and data in slack space—which may be critical to the investigative process. Once data from the source are altered, they may no longer be admissible in court. Continuing the investigation, documenting the date, time and data altered, are actions that may not be admissible in legal proceedings. The organization would need to know the details of collecting and preserving forensic evidence relevant to their jurisdiction.

S4-84   Which of the following recovery strategies has the **GREATEST** chance of failure?

A. Hot site
B. Redundant site
C. Reciprocal arrangement
D. Cold site

C   A reciprocal arrangement is an agreement that allows two organizations to back up each other during a disaster. This approach sounds desirable, but has the greatest chance of failure due to problems in keeping agreements and plans up to date. A hot site is incorrect because it is a site kept fully equipped with processing capabilities and other services by the vendor. A redundant site is incorrect because it is a site equipped and configured exactly like the primary site. A cold site is incorrect because it is a building having a basic environment such as electrical wiring, air conditioning, flooring, etc. and is ready to receive equipment in order to operate.

S4-85   Recovery point objectives (RPOs) can be used to determine which of the following?

    A. Maximum tolerable period of data loss
    B. Maximum tolerable downtime
    C. Baseline for operational resiliency
    D. Time to restore backups

A   The RPO is determined based on the acceptable data loss in the case of disruption of operations. It indicates the farthest point in time prior to the incident to which it is acceptable to recover the data. RPO effectively quantifies the permissible amount of data loss in the case of interruption. It also dictates the frequency of backups required for a given data set since the smaller the allowable gap in data, the more frequent that backups must occur.

S4-86   Which of the following disaster recovery testing techniques is the **MOST** cost-effective way to determine the effectiveness of the plan?

    A. Preparedness tests
    B. Paper tests
    C. Full operational tests
    D. Actual service disruption

A   Preparedness tests would involve simulation of the entire test in phases and help the team better understand and prepare for the actual test scenario. Choices B, C and D are not cost-effective ways to establish plan effectiveness. Paper tests in a walk-through do not include simulation and so there is less learning and it is difficult to obtain evidence that the team has understood the test plan. Choice D is not recommended in most cases. Choice C would require an approval from management, is not easy or practical to test in most scenarios and may itself trigger a disaster.

S4-87   When electronically stored information is requested during a fraud investigation, which of the following should be the **FIRST** priority?

    A. Assigning responsibility for acquiring the data
    B. Locating the data and preserving the integrity of the data
    C. Creating a forensically sound image
    D. Issuing a litigation hold to all affected parties

B   Locating the data and preserving data integrity is the only correct answer because it represents the primary responsibility of an investigator and is a complete and accurate statement of the first priority. While assigning responsibility for acquiring the data is a step that should be taken, it is not the first step or the highest priority. Creating a forensically sound image may or may not be a necessary step, depending on the type of investigation, but it would never be the first priority. Issuing a litigation hold to all affected parties might be a necessary step early on in an investigation of certain types, but not the first priority.

S4-88   When creating a forensic image of a hard drive, which of the following should be the **FIRST** step?

A. Identify a recognized forensics software tool to create the image.
B. Establish a chain of custody log.
C. Connect the hard drive to a write blocker.
D. Generate a cryptographic hash of the hard drive contents.

B   The first step in any investigation requiring the creation of a forensic image should always be to maintain the chain of custody. Identifying a recognized forensics software tool to create the image is one of the important steps, but it should come after several of the other options. Connecting the hard drive to a write blocker is an important step, but it must be done after the chain of custody has been established. Generating a cryptographic hash of the hard drive contents is another important step, but one that comes after several of the other options.

S4-89   When a significant security breach occurs, what should be reported **FIRST** to senior management?

A. A summary of the security logs that illustrates the sequence of events
B. An explanation of the incident and corrective action taken
C. An analysis of the impact of similar attacks at other organizations
D. A business case for implementing stronger logical access controls

B   When reporting an incident to senior management, the initial information to be communicated should include an explanation of what happened and how the breach was resolved. A summary of security logs would be too technical to report to senior management. An analysis of the impact of similar attacks and a business case for improving controls would be desirable; however, these would be communicated later in the process.

S4-90   The **BEST** time to determine who should be responsible for declaring a disaster is:

A. during the establishment of the plan.
B. once an incident has been confirmed by operations staff.
C. after fully testing the incident management plan.
D. after the implementation details of the plan have been approved.

A   Roles and responsibilities for all involved in incident response should be established when the incident response plan is established. Determining roles and responsibilities during a disaster is not the best time to make such decisions, unless it is absolutely necessary. While testing the plan may drive some changes in roles based on test results, roles (including who declares the disaster) should have been established before testing and plan approval.

S4-91   The **PRIMARY** objective of incident response is to:

A. investigate and report results of the incident to management.
B. gather evidence.
C. minimize business disruptions.
D. assist law enforcement in investigations.

C   The primary role of incident response is to detect, respond to and contain incidents so that impact to business operations is minimized. Choice A is a responsibility of incident response teams, but not the primary objective. Choices B and D are activities that an incident response team may conduct, depending on circumstances, but neither is a primary objective.

S4-92   An information security manager is in the process of investigating a network intrusion. One of the
         enterprise's employees is a suspect. The manager has just obtained the suspect's computer and hard drive.
         Which of the following is the **BEST** next step?

         A. Create an image of the hard drive.
         B. Encrypt the data on the hard drive.
         C. Examine the original hard drive.
         D. Create a logical copy of the hard drive.

A   One of the first steps in an investigation is to create an image of the original hard drive. A physical copy
         will copy the data, block by block, including any hidden data blocks and hidden partitions that can be used
         to conceal evidence. Encryption is not required. Examining the hard drive is not good practice. A logical
         copy will only copy the files and folders and may not copy the necessary data to properly examine the hard
         drive for forensic evidence.

S4-93   The factor that is **MOST** likely to result in identification of security incidents is:

         A. effective communication and reporting processes.
         B. clear policies detailing incident severity levels.
         C. intrusion detection system (IDS) capabilities.
         D. security awareness training.

D   Ensuring that employees have the knowledge to recognize and report a suspected incident is most likely
         to result in identification of security incidents. Timely communication and reporting is only useful once
         identification of an incident has occurred. Understanding how to establish severity levels is important, but
         not the essential element of ensuring that the information security manager is aware of anomalous events
         that might signal an incident. IDSs are useful for detecting IT-related incidents, but not useful in identifying
         other types of incidents such as social engineering or physical intrusion.

S4-94   Which of the following functions is responsible for determining the members of the enterprise's
         response teams?

         A. Governance
         B. Risk management
         C. Compliance
         D. Information security

D   The information security manager, or designated manager for incident response, should select the team
         members required to ensure that all required disciplines are represented on the team. The governance
         function will determine the strategy and policies that will set the scope and charter for incident management
         and response capabilities. While response is a component of managing risk, the basis for risk management
         is determined by governance and strategy requirements. Compliance would not be directly related to this
         activity, although this function may have representation on the incident response team.

S4-95   The typical requirement for security incidents to be resolved quickly and service restored is:

    A.   always the best option for an enterprise.
    B.   often in conflict with effective problem management.
    C.   the basis for enterprise risk management (ERM) activities.
    D.   a component of forensics training.

B   Problem management is focused on investigating and uncovering the root cause of incidents, which will often be a problem when restoring service compromises the evidence needed. Quickly restoring service will not always be the best option such as in cases of criminal activity, which requires preservation of evidence precluding use of the systems involved. Managing risk goes beyond the quick restoration of services, e.g., if doing so increased some other risk disproportionately. Forensics is concerned with legally adequate collection and preservation of evidence, not with service continuity.

S4-96   Which of the following should be the **FIRST** action to take when a fire spreads throughout the building?

    A.   Check the facility access logs.
    B.   Call together the crisis management team.
    C.   Launch the disaster recovery plan (DRP).
    D.   Launch the business continuity plan (BCP).

A   Safety of people always comes first; therefore, verifying access logs of personnel to the facility should be the first action in order to ensure that all staff can be accounted for. Calling the crisis management team together should be done after the initial emergency response (i.e., evacuation of people). Launching the DRP is not the first action. Launching the BCP is not the first action.

S4-97   Which of the following tests gives the **MOST** assurance that a business continuity plan (BCP) works, without potentially impacting business operations?

    A.   Checklist tests
    B.   Simulation tests.
    C.   Walk-through tests
    D.   Full operational tests

B   Business continuity coordinators come together to practice executing a plan based on a specific scenario. This does not interrupt normal operations and provides the most assurance of the given nonintrusive methods. With checklist tests, copies of the BCP are distributed to various persons for review. In these tests, people do not exercise a plan. In walk-through tests, representatives come together to go over the plan (one or more scenarios) and ensure the plan's accuracy. The plan itself is not executed. Full operational tests are the most intrusive to regular operations and business productivity. The original site is actually shut down and processing is performed at another site, thus providing the most assurance, but interrupting normal business productivity.

S4-98   An employee's computer has been infected with a new virus. What should be the **FIRST** action?

    A.   Execute the virus scan.
    B.   Report the incident to senior management.
    C.   Format the hard disk.
    D.   Disconnect the computer from the network.

D   The first action should be containing the risk, i.e., disconnecting the computer so that it will not infect other computers on the network. The virus may start infecting other computers while the virus scan is running. Only when the impact to the IT environment is significant should it be reported to senior management. A case of virus infection does not warrant the action. Formatting the hard disk is the last resort.

S4-99   The **PRIMARY** reason for senior management review of information security incidents is to:

A. ensure adequate corrective actions were implemented.
B. demonstrate management commitment to the information security process.
C. evaluate the incident response process for deficiencies.
D. evaluate the ability of the security team.

A   Although some corrective actions are being taken by the security team and the incident response team, management review will ensure whether there are any other corrective actions that need to be taken. Sometimes this will result in improvements to information security policies. Management will not review information security incidents merely to demonstrate management commitment. Management will not perform a review for fault finding such as examining the incidence response process for deficiencies and the ability of the security team.

S4-100   Observations made by staff during a disaster recovery test are **PRIMARILY** reviewed to:

A. identify people who have not followed the process.
B. determine lessons learned.
C. identify equipment that is needed.
D. maintain evidence of review.

B   After a test, results should be reviewed to ensure that lessons learned are applied. It is not the aim of observation to identify people who have not followed the process. Identifying equipment that is needed may be part of the lessons learned, but is not the sole reason for the review. Review is conducted not only to maintain evidence, but to make improvements.

S4-101   The **PRIMARY** selection criterion for an offsite media storage facility is:

A. that the primary and offsite facilities not be subject to the same environmental disasters.
B. that the offsite storage facility be in close proximity to the primary site.
C. the overall storage and maintenance costs of the offsite facility.
D. the availability of cost-effective media transportation services.

A   It is important to prevent a disaster that could affect both sites. The distance between sites may be important in cases of widespread disasters; however, this is covered by choice A. The costs should not be the primary criteria to selection. A cost-effective media transport service may be a consideration, but is not the main concern.

S4-102   Security-related breaches are assessed and contained through:

A. disaster recovery.
B. incident response.
C. a forensic analysis.
D. the IT support team.

B   The incident response plan must be activated when an incident occurs. A disaster recovery plan (DRP) can be activated as part of an incident response plan (IRP). A forensic analysis can be part of an IRP, but is not necessarily a component. IT support can be part of a response team, but the team can have other members.

S4-103   The recovery time objective (RTO) is reached at which of the following milestones?

    A. Disaster declaration
    B. Recovery of the backups
    C. Restoration of the system
    D. Return to business as usual processing

C   The recovery time objective (RTO) is based on the amount of time required to restore a system; disaster declaration occurs at the beginning of this period. Recovery of the backups occurs shortly after the beginning of this period. Return to business as usual processing occurs significantly later than the RTO. RTO is an "objective," and full restoration may or may not coincide with the RTO. RTO can be the minimum acceptable operational level, far short of normal operations.

S4-104   The recovery point objective (RPO) requires which of the following?

    A. Disaster declaration
    B. Before-image restoration
    C. System restoration
    D. After-image processing

B   The recovery point objective (RPO) is the point in the processing flow at which system recovery should occur. This is the predetermined state of the application processing and data used to restore the system and to continue the processing flow. Disaster declaration is independent of this processing checkpoint. Restoration of the system can occur at a later date, as does the return to normal, after-image processing.

S4-105   Who would be in the **BEST** position to determine the recovery point objective (RPO) for business applications?

    A. Business continuity coordinator
    B. Chief operations officer (COO)
    C. Information security manager
    D. Internal audit

B   The recovery point objective (RPO) is the processing checkpoint to which systems are recovered. In addition to data owners, the chief operations officer (COO) is the most knowledgeable person to make this decision. It would be inappropriate for the information security manager or an internal audit to determine the RPO because they are not directly responsible for the data or the operation.

S4-106   When the computer incident response team (CIRT) finds clear evidence that a hacker has penetrated the corporate network and modified customer information, an information security manager should **FIRST** notify:

    A. the information security steering committee.
    B. customers who may be impacted.
    C. data owners who may be impacted.
    D. regulatory agencies overseeing privacy.

C   The data owners should be notified first so they can take steps to determine the extent of the damage and coordinate a plan for corrective action with the computer incident response team. Other parties will be notified later as required by corporate policy and regulatory requirements.

S4-107   The systems administrator did not immediately notify the security officer about a malicious attack.
          An information security manager could prevent this situation by:

          A. periodically testing the incident response plans.
          B. regularly testing the intrusion detection system (IDS).
          C. establishing mandatory training of all personnel.
          D. periodically reviewing incident response procedures.

A   Security incident response plans should be tested to find any deficiencies and improve existing processes.
          Testing the intrusion detection system (IDS) is a good practice but would not have prevented this situation.
          All personnel need to go through formal training to ensure that they understand the process, tools and
          methodology involved in handling security incidents. However, testing of the actual plans is more effective
          in ensuring the process works as intended. Reviewing the response procedures is not enough; the security
          response plan needs to be tested on a regular basis.

S4-108   Which of the following would a security manager establish to determine the target for restoration of
          normal processing?

          A. Recovery time objective (RTO)
          B. Maximum tolerable outage (MTO)
          C. Recovery point objectives (RPOs)
          D. Services delivery objectives (SDOs)

A   Recovery time objective (RTO) is the length of time from the moment of an interruption until the time
          the process must be functioning at a service level sufficient to limit financial and operational impacts to
          an acceptable level. Maximum tolerable outage (MTO) is the maximum time for which an organization
          can operate in a reduced mode. Recovery point objectives (RPOs) relate to the age of the data required for
          recovery. Services delivery objectives (SDOs) are the levels of service required in reduced mode.

S4-109   Which of the following should be the **PRIMARY** basis for making a decision to establish an alternate site
          for disaster recovery?

          A. A business impact analysis (BIA), which identifies the requirements for continuous availability of
              critical business processes
          B. Adequate distance between the primary site and the alternate site so that the same disaster does not
              simultaneously impact both sites
          C. A benchmarking analysis of similarly situated enterprises in the same geographic region to demonstrate
              due diligence
          D. Differences between the regulatory requirements applicable at the primary site and those at the
              alternate site

A   The BIA will help determine the recovery time objective (RTO) and recovery point objective (RPO) for the
          enterprise. This information will drive the decision on the appropriate level of protection for its assets. Natural
          disasters and regulatory requirements are just two of many factors that an enterprise must consider when it
          decides whether to pursue an alternate site for disaster recovery. While a benchmark could provide useful
          information, the decision should be based on a BIA, which considers factors specific to the enterprise.

S4-110   During a business continuity plan (BCP) test, one department discovered that its new software application was not going to be restored soon enough to meet the needs of the business. This situation can be avoided in the future by:

A. conducting a periodic and event-driven business impact analysis (BIA) to determine the needs of the business during a recovery.
B. assigning new applications a higher degree of importance and scheduling them for recovery first.
C. developing a help-desk ticket process that allows departments to request recovery of software during a disaster.
D. conducting a thorough risk assessment prior to purchasing the software.

A   A periodic BIA can help compensate for changes in the needs of the business for recovery during a disaster. Choice B is an incorrect assumption regarding the automatic importance of a new program. Choice C is not an appropriate recovery procedure because it allows individual business units to make unilateral decisions without consideration of broader implications. The risk assessment may not include the BIA.

S4-111   The main mail server of a financial institution has been compromised at the superuser level; the only way to ensure the system is secure would be to:

A. change the root password of the system.
B. implement multifactor authentication.
C. rebuild the system from the original installation medium.
D. disconnect the mail server from the network.

C   Rebuilding the system from the original installation medium is the only way to ensure all security vulnerabilities and potential stealth malicious programs have been destroyed. Changing the root password of the system does not ensure the integrity of the mail server. Implementing multifactor authentication is an aftermeasure and does not clear existing security threats. Disconnecting the mail server from the network is an initial step, but does not guarantee security.

S4-112   Which of the following would present the **GREATEST** risk to information security?

A. Virus signature files updates are applied to all servers every day
B. Security access logs are reviewed within five business days
C. Critical patches are applied within 24 hours of their release
D. Security incidents are investigated within five business days

D   Security incidents are configured to capture system events that are important from the security perspective; they include incidents also captured in the security access logs and other monitoring tools. Although, in some instances, they could wait for a few days before they are researched, from the options given this would have the greatest risk to security. Most often, they should be analyzed as soon as possible. Virus signatures should be updated as often as they become available by the vendor, while critical patches should be installed as soon as they are reviewed and tested, which could occur in 24 hours.

S4-113   Which of the following is the **MOST** important area of focus when examining potential security compromise of a new wireless network?

    A. Signal strength
    B. Number of administrators
    C. Bandwidth
    D. Encryption strength

B   The number of individuals with access to the network configuration presents a security risk. Encryption strength is an area where wireless networks tend to fall short; however, the potential to compromise the entire network is higher when an inappropriate number of people can alter the configuration. Signal strength and network bandwidth are secondary issues.

S4-114   When security policies are strictly enforced, the initial impact is that:

    A. they may have to be modified more frequently.
    B. they will be less subject to challenge.
    C. the total cost of security is increased.
    D. the need for compliance reviews is decreased.

C   When security policies are strictly enforced, more resources are initially required, thereby increasing the total cost of security. There would be less need for frequent modification. Challenges would be rare and the need for compliance reviews would not necessarily be less.

S4-115   Which of the following is the **BEST** indicator that operational risks are effectively managed in an enterprise?

    A. A tested business continuity/disaster recovery plan (BCP/DRP)
    B. An increase in timely reporting of incidents by employees
    C. Extent of risk management education
    D. Regular review of risks by senior management

A   A tested BCP/DRP is the best indicator that operational risks are managed effectively in the enterprise. Reporting incidents by employees is an indicator, but not the best choice because it is dependent upon the knowledge of the employees. Extent of risk management education is not correct since this may not necessarily indicate that risks are effectively managed in the enterprise. A high level of risk management education would help, but would not necessarily mean that risks are managed effectively. Regular review of risks by senior management is not correct since this may not necessarily indicate that risks are effectively managed in the enterprise. Top management involvement would greatly help, but would not necessarily mean that risks are managed effectively.

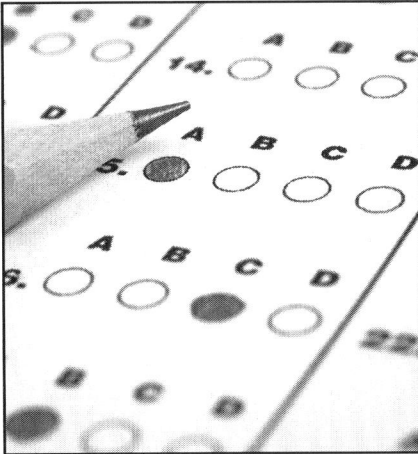

POSTTEST

If you wish to take a posttest to determine strengths and weaknesses, the Sample Exam begins on page 195 and the posttest answer sheet begins on page 229. You can score your posttest with the Sample Exam Answer and Reference Key on page 225.

Page intentionally left blank

SAMPLE EXAM

1. In a large enterprise, an information security awareness program will be **MOST** effective if it is:

    A. developed by a professional training company.
    B. embedded into the orientation process.
    C. customized to the audience using the appropriate delivery channel.
    D. required by the information security policy.

2. Which of the following would be the **MOST** relevant factor when defining the information classification policy?

    A. Quantity of information
    B. Available IT infrastructure
    C. Benchmarking
    D. Requirements of data owners

3. An information security manager has been assigned to implement more restrictive preventive controls. By doing so, the net effect will be to **PRIMARILY** reduce the:

    A. threat.
    B. loss.
    C. vulnerability.
    D. probability.

4. Which of the following is the **MOST** appropriate position to sponsor the design and implementation of a new security infrastructure in a large global enterprise?

    A. Chief security officer (CSO)
    B. Chief operating officer (COO)
    C. Chief privacy officer (CPO)
    D. Chief legal counsel (CLC)

5. The **PRIORITY** action to be taken when a server is infected with a virus is to:

    A. isolate the infected server(s) from the network.
    B. identify all potential damage caused by the infection.
    C. ensure that the virus database files are current.
    D. establish security weaknesses in the firewall.

6. Senior management commitment and support for information security will **BEST** be attained by an information security manager by emphasizing:

    A. organizational risk.
    B. organizationwide metrics.
    C. security needs.
    D. the responsibilities of organizational units.

7. The service level agreement (SLA) for an outsourced IT function does not reflect an adequate level of protection. In this situation an information security manager should:

    A. ensure the provider is made liable for losses.
    B. recommend not renewing the contract upon expiration.
    C. recommend the immediate termination of the contract.
    D. determine the current level of security.

8.   When performing a business impact analysis (BIA), which of the following should calculate the recovery time and cost estimates?

   A. Business continuity coordinator
   B. Information security manager
   C. Business process owners
   D. IT management

9.   The **PRIMARY** objective when selecting controls and countermeasures is to:

   A. protect against all threats.
   B. reduce costs.
   C. optimize protection and usability.
   D. restrict employee access.

10.   Who in an organization has the responsibility for classifying information?

   A. Data custodian
   B. Database administrator
   C. Information security officer
   D. Data owner

11.   Security-related breaches are assessed and contained through:

   A. disaster recovery.
   B. incident response.
   C. a forensic analysis.
   D. the IT support team.

12.   Which of the following attacks is **BEST** mitigated by utilizing strong passwords?

   A. Man-in-the-middle attack
   B. Brute force attack
   C. Remote buffer overflow
   D. Root kit

13.   An outsource service provider must handle sensitive customer information. Which of the following is **MOST** important for an information security manager to know?

   A. Security in storage and transmission of sensitive data
   B. Provider's level of compliance with industry standards
   C. Security technologies in place at the facility
   D. Results of the latest independent security review

14.   Generally, who should determine the classification of an information asset?

   A. The asset custodian
   B. The security manager
   C. Senior management
   D. The asset owner

15.   Investment in security technology and processes should be based on:

        A. clear alignment with the goals and objectives of the organization.
        B. success cases that have been experienced in previous projects.
        C. best business practices.
        D. safeguards that are inherent in existing technology.

16.   To achieve effective strategic alignment of security initiatives, it is important that:

        A. steering committee leadership be selected by rotation.
        B. inputs be obtained and consensus achieved between the major organizational units.
        C. the business strategy be updated periodically.
        D. procedures and standards be approved by all departmental heads.

17.   Which of the following is **MOST** essential for a risk management program to be effective?

        A. Flexible security budget
        B. Sound risk baseline
        C. Detection of new risk
        D. Accurate risk reporting

18.   The **PRIMARY** objective of incident response is to:

        A. investigate and report results of the incident to management.
        B. gather evidence.
        C. minimize business disruptions.
        D. assist law enforcement in investigations.

19.   Recovery point objectives (RPOs) can be used to determine which of the following?

        A. Maximum tolerable period of data loss
        B. Maximum tolerable downtime
        C. Baseline for operational resiliency
        D. Time to restore backups

20.   Which of the following are the essential ingredients of a business impact analysis (BIA)?

        A. Downtime tolerance, resources and criticality
        B. Cost of business outages in a year as a factor of the security budget
        C. Business continuity testing methodology being deployed
        D. Structure of the crisis management team

21.   Which of the following is responsible for legal and regulatory liability?

        A. Chief security officer (CSO)
        B. Chief legal counsel (CLC)
        C. Board and senior management
        D. Information security steering group

22.   The decision as to whether a risk has been reduced to an acceptable level should be determined by:

        A. organizational requirements.
        B. information systems requirements.
        C. information security requirements.
        D. international standards.

23.   The factor that is **MOST** likely to result in identification of security incidents is:

        A.   effective communication and reporting processes.
        B.   clear policies detailing incident severity levels.
        C.   intrusion detection system (IDS) capabilities.
        D.   security awareness training.

24.   The valuation of IT assets should be performed by:

        A.   an IT security manager.
        B.   an independent security consultant.
        C.   the chief financial officer (CFO).
        D.   the information owner.

25.   An effective risk management program should reduce risk to:

        A.   zero.
        B.   an acceptable level.
        C.   an acceptable percent of revenue.
        D.   an acceptable probability of occurrence.

26.   Why is "slack space" of value to an information security manager as part of an incident investigation?

        A.   Hidden data may be stored there
        B.   The slack space contains login information
        C.   Slack space is encrypted
        D.   It provides flexible space for the investigation

27.   Which of the following is an inherent weakness of signature-based intrusion detection systems?

        A.   A higher number of false positives
        B.   New attack methods will be missed
        C.   Long duration probing will be missed
        D.   Attack profiles can be easily spoofed

28.   Which of the following is the **MOST** important action to take when engaging third-party consultants to conduct an attack and penetration test?

        A.   Request a list of the software to be used
        B.   Provide clear directions to IT staff
        C.   Monitor intrusion detection system (IDS) and firewall logs closely
        D.   Establish clear rules of engagement

29.   A common concern with poorly written web applications is that they can allow an attacker to:

        A.   gain control through a buffer overflow.
        B.   conduct a distributed denial of service (DoS) attack.
        C.   abuse a race condition.
        D.   inject structured query language (SQL) statements.

30. When a major vulnerability in the security of a critical web server is discovered, immediate notification should be made to the:

    A. system owner to take corrective action.
    B. incident response team to investigate.
    C. data owners to mitigate damage.
    D. development team to remediate.

31. The data access requirements for an application should be determined by the:

    A. legal department.
    B. compliance officer.
    C. information security manager.
    D. business owner.

32. Which of the following is the **BEST** way to erase confidential information stored on magnetic tapes?

    A. Performing a low-level format
    B. Rewriting with zeros
    C. Burning them
    D. Degaussing them

33. Which of the following groups would be in the **BEST** position to perform a risk analysis for a business?

    A. External auditors
    B. A peer group within a similar business
    C. Process owners
    D. A specialized management consultant

34. An incident response policy must contain:

    A. updated call trees.
    B. escalation criteria.
    C. press release templates.
    D. critical backup files inventory.

35. The **BEST** way to justify the implementation of a single sign-on (SSO) product is to use:

    A. return on investment (ROI).
    B. a vulnerability assessment.
    C. annual loss expectancy (ALE).
    D. a business case.

36. An intrusion detection system (IDS) should:

    A. run continuously
    B. ignore anomalies
    C. require a stable, rarely changed environment
    D. be located on the network

37.   An enterprise requires the use of Windows XP Service Pack 3 version on all desktops and Windows 2003
        Service Pack 1 version on all servers. This is an example of a:

        A. policy.
        B. guideline.
        C. standard.
        D. procedure.

38.   In designing a backup strategy that will be consistent with a disaster recovery strategy, the **PRIMARY**
        factor to be taken into account will be the:

        A. volume of sensitive data.
        B. recovery point objective (RPO).
        C. recovery time objective (RTO).
        D. interruption window.

39.   The **BEST** way to determine if an anomaly-based intrusion detection system (IDS) is properly installed is to:

        A. simulate an attack and review IDS performance.
        B. use a honeypot to check for unusual activity.
        C. audit the configuration of the IDS.
        D. benchmark the IDS against a peer site.

40.   Which of the following functions is responsible for determining the members of the enterprise's
        response teams?

        A. Governance
        B. Risk management
        C. Compliance
        D. Information security

41.   An information security manager mapping a job description to types of data access is **MOST** likely to
        adhere to which of the following information security principles?

        A. Ethics
        B. Proportionality
        C. Integration
        D. Accountability

42.   Which of the following is **MOST** likely to be discretionary?

        A. Policies
        B. Procedures
        C. Guidelines
        D. Standards

43.   Which of the following would be the **BEST** option to improve accountability for a system administrator
        who has security functions?

        A. Include security responsibilities in the job description
        B. Require the administrator to obtain security certification
        C. Train the system administrator on penetration testing and vulnerability assessment
        D. Train the system administrator on risk assessment

44.   Which of the following is the **PRIMARY** prerequisite to implementing data classification within
       an organization?

       A.   Defining job roles
       B.   Performing a risk assessment
       C.   Identifying data owners
       D.   Establishing data retention policies

45.   Which of the following would help management determine the resources needed to mitigate a risk to
       the organization?

       A.   Risk analysis process
       B.   Business impact analysis (BIA)
       C.   Risk management balanced scorecard
       D.   Risk-based audit program

46.   Which of the following **BEST** ensures that modifications made to in-house developed business applications
       do not introduce new security exposures?

       A.   Stress testing
       B.   Patch management
       C.   Change management
       D.   Security baselines

47.   Which of the following is the **MOST** important requirement for setting up an information security
       infrastructure for a new system?

       A.   Performing a business impact analysis (BIA)
       B.   Considering personal information devices as part of the security policy
       C.   Initiating IT security training and familiarization
       D.   Basing the information security infrastructure on risk assessment

48.   What is the **GREATEST** advantage of documented guidelines and operating procedures from a
       security perspective?

       A.   Provide detailed instructions on how to carry out different types of tasks
       B.   Ensure consistency of activities to provide a more stable environment
       C.   Ensure compliance to security standards and regulatory requirements
       D.   Ensure reusability to meet compliance to quality requirements

49.   When an organization is using an automated tool to manage and house its business continuity plans, which
       of the following is the **PRIMARY** concern?

       A.   Ensuring accessibility should a disaster occur
       B.   Versioning control as plans are modified
       C.   Broken hyperlinks to resources stored elsewhere
       D.   Tracking changes in personnel and plan assets

50.   Which of the following is **MOST** important in developing a security strategy?

       A.   Creating a positive business security environment
       B.   Understanding key business objectives
       C.   Having a reporting line to senior management
       D.   Allocating sufficient resources to information security

51.   Which of the following individuals would be in the **BEST** position to sponsor the creation of an information security steering group?

A. Information security manager
B. Chief operating officer (COO)
C. Internal auditor
D. Legal counsel

52.   The impact of losing frame relay network connectivity for 18-24 hours should be calculated using the:

A. hourly billing rate charged by the carrier.
B. value of the data transmitted over the network.
C. aggregate compensation of all affected business users.
D. financial losses incurred by affected business units.

53.   The **PRIMARY** objective of a security steering group is to:

A. ensure information security covers all business functions.
B. ensure information security aligns with business goals.
C. raise information security awareness across the organization.
D. implement all decisions on security management across the organization.

54.   Which of the following is **MOST** important in determining whether a disaster recovery test is successful?

A. Only business data files from offsite storage are used
B. IT staff fully recovers the processing infrastructure
C. Critical business processes are duplicated
D. All systems are restored within recovery time objectives (RTOs)

55.   A business impact analysis (BIA) is the **BEST** tool for calculating:

A. total cost of ownership.
B. priority of restoration.
C. annualized loss expectancy (ALE).
D. residual risk.

56.   Which of the following actions should be taken when an online trading company discovers a network attack in progress?

A. Shut off all network access points
B. Dump all event logs to removable media
C. Isolate the affected network segment
D. Enable trace logging on all event

57.   The **PRIMARY** reason for senior management review of information security incidents is to:

A. ensure adequate corrective actions were implemented.
B. demonstrate management commitment to the information security process.
C. evaluate the incident response process for deficiencies.
D. evaluate the ability of the security team.

CISM Review Questions, Answers & Explanations Manual 2012
ISACA. All Rights Reserved.

58. During which phase of development is it **MOST** appropriate to begin assessing the risk of a new application system?

    A. Feasibility
    B. Design
    C. Development
    D. Testing

59. Which of the following would be of **GREATEST** importance to the security manager in determining whether to further mitigate residual risk?

    A. Historical cost of the asset
    B. Acceptable level of potential business impacts
    C. Cost versus benefit of additional mitigating controls
    D. Annualized loss expectancy (ALE)

60. Which of the following would typically be the **MOST** effective physical security access control for the main entrance to a data center?

    A. Mantrap
    B. Biometric lock
    C. Closed-circuit television (CCTV)
    D. Security guard

61. When personal information is transmitted across networks, there MUST be adequate controls over:

    A. change management.
    B. privacy protection.
    C. consent to data transfer.
    D. encryption devices.

62. The **BEST** approach in managing a security incident involving a successful penetration should be to:

    A. allow business processes to continue during the response.
    B. allow the security team to assess the attack profile.
    C. permit the incident to continue to trace the source.
    D. examine the incident response process for deficiencies.

63. Which of the following is the **MOST** appropriate use of gap analysis?

    A. Evaluating a business impact analysis (BIA)
    B. Developing a balanced business scorecard
    C. Demonstrating the relationship between controls
    D. Measuring current state vs. desired future state

64. Good information security procedures should:

    A. define the allowable limits of behavior.
    B. underline the importance of security governance.
    C. describe security baselines for each platform.
    D. be updated frequently as new software is released.

65. An enterprise has a network of suppliers that it allows to remotely access an important database that contains critical supply chain data. What is the **BEST** control to ensure that the individual supplier representatives who have access to the system do not improperly access or modify information within this system?

A. User access rights
B. Biometric access controls
C. Password authentication
D. Two-factor authentication

66. A message that has been encrypted by the sender's private key and again by the receiver's public key achieves:

A. authentication and authorization.
B. confidentiality and integrity.
C. confidentiality and nonrepudiation.
D. authentication and nonrepudiation.

67. While implementing information security governance an organization should **FIRST**:

A. adopt security standards.
B. determine security baselines.
C. define the security strategy.
D. establish security policies.

68. Of the following, the **BEST** method for ensuring that temporary employees do not receive excessive access rights is:

A. mandatory access controls.
B. discretionary access controls.
C. lattice-based access controls.
D. role-based access controls.

69. Security awareness training is **MOST** likely to lead to which of the following?

A. Decrease in intrusion incidents
B. Increase in reported incidents
C. Decrease in security policy changes
D. Increase in access rule violations

70. The **MOST** important factor in ensuring the success of an information security program is effective:

A. communication of information security requirements to all users in the organization.
B. formulation of policies and procedures for information security.
C. alignment with organizational goals and objectives .
D. monitoring compliance with information security policies and procedures.

71. Which of the following should be included in an annual information security budget that is submitted for management approval?

A. A cost-benefit analysis of budgeted resources
B. All of the resources that are recommended by the business
C. Total cost of ownership (TCO)
D. Baseline comparisons

72. Which of the following steps in conducting a risk assessment should be performed **FIRST**?

    A. Identify business assets
    B. Identify business risks
    C. Assess vulnerabilities
    D. Evaluate key controls

73. Which of the following should be performed **EXCLUSIVELY** by the information security department?

    A. Monitoring unauthorized access to operating systems
    B. Configuring user access to operating systems
    C. Approving operating system access standards
    D. Configuring the firewall to protect operating systems

74. A possible breach of an organization's IT system is reported by the project manager. What is the **FIRST** thing the incident response manager should do?

    A. Run a port scan on the system
    B. Disable the logon ID
    C. Investigate the system logs
    D. Validate the incident

75. Which of the following risks would **BEST** be assessed using qualitative risk assessment techniques?

    A. Theft of purchased software
    B. Power outage lasting 24 hours
    C. Permanent decline in customer confidence
    D. Temporary loss of e-mail due to a virus attack

76. Observations made by staff during a disaster recovery test are **PRIMARILY** reviewed to:

    A. identify people who have not followed the process.
    B. determine lessons learned.
    C. identify equipment that is needed.
    D. maintain evidence of review.

77. Nonrepudiation can **BEST** be assured by using:

    A. delivery path tracing.
    B. reverse lookup translation.
    C. out-of-band channels.
    D. digital signatures.

78. An organization has to comply with recently published industry regulatory requirements—compliance that potentially has high implementation costs. What should the information security manager do **FIRST**?

    A. Implement a security committee.
    B. Perform a gap analysis.
    C. Implement compensating controls.
    D. Demand immediate compliance.

79. The **PRIMARY** purpose of using risk analysis within a security program is to:

   A. justify the security expenditure.
   B. help businesses prioritize the assets to be protected.
   C. inform executive management of residual risk value.
   D. assess exposures and plan remediation.

80. Risk management programs are designed to reduce risk to:

   A. a level that is too small to be measurable.
   B. the point at which the benefit exceeds the expense.
   C. a level that the organization is willing to accept.
   D. a rate of return that equals the current cost of capital.

81. To help ensure that contract personnel do not obtain unauthorized access to sensitive information, an information security manager should **PRIMARILY**:

   A. set their accounts to expire in six months or less.
   B. avoid granting system administration roles.
   C. ensure they successfully pass background checks.
   D. ensure their access is approved by the data owner.

82. In order to highlight to management the importance of network security, the security manager should **FIRST**:

   A. develop a security architecture.
   B. install a network intrusion detection system (NIDS) and prepare a list of attacks.
   C. develop a network security policy.
   D. conduct a risk assessment.

83. An information security manager is in the process of investigating a network intrusion. One of the enterprise's employees is a suspect. The manager has just obtained the suspect's computer and hard drive. Which of the following is the **BEST** next step?

   A. Create an image of the hard drive.
   B. Encrypt the data on the hard drive.
   C. Examine the original hard drive.
   D. Create a logical copy of the hard drive.

84. Which of the following is the **MOST** effective at preventing an unauthorized individual from following an authorized person through a secured entrance (tailgating or piggybacking)?

   A. Card-key door locks
   B. Photo identification
   C. Biometric scanners
   D. Awareness training

85. Which of the following is characteristic of decentralized information security management across a geographically dispersed organization?

   A. More uniformity in quality of service
   B. Better adherence to policies
   C. Better alignment to business unit needs
   D. More savings in total operating costs

86.   A mission-critical system has been identified as having an administrative system account with attributes that prevent locking and change of privileges and name. Which would be the **BEST** approach to prevent successful brute forcing of the account?

    A. Prevent the system from being accessed remotely
    B   Create a strong random password
    C. Ask for a vendor patch
    D. Track usage of the account by audit trails

87.   Which of the following is the **MOST** effective security measure to protect data held on mobile computing devices?

    A. Biometric access control
    B. Encryption of stored data
    C. Power-on passwords
    D. Protection of data being transmitted

88.   Which of the following would govern which information assets need more protection than other information assets?

    A. A data custodian
    B. Asset identification
    C. Data classification
    D. An acceptable use policy

89.   Information security governance is **PRIMARILY** driven by:

    A. technology constraints.
    B. regulatory requirements.
    C. litigation potential.
    D. business strategy.

90.   Which program element should be implemented **FIRST** in asset classification and control?

    A. Risk assessment
    B. Classification
    C. Valuation
    D. Risk mitigation

91.   Simple Network Management Protocol v2 (SNMP v2) is used frequently to monitor networks. Which of the following vulnerabilities does it always introduce?

    A. Remote buffer overflow
    B. Cross site scripting
    C. Clear text authentication
    D. Man-in-the-middle attack

92.   Which of the following factors will **MOST** affect the extent to which controls should be layered?

    A. The extent to which controls are procedural
    B. Controls subject to the same threat
    C. The maintenance cost of controls
    D. Controls that fail in a closed condition

93. Data owners are normally responsible for which of the following?

    A. Applying emergency changes to application data
    B. Administering security over database records
    C. Migrating application code changes to production
    D. Determining the level of application security required

94. A global financial institution has decided not to take any further action on a denial of service (DoS) risk found by the risk assessment team. The **MOST** likely reason they made this decision is that:

    A. there are sufficient safeguards in place to prevent this risk from happening.
    B. the needed countermeasure is too complicated to deploy.
    C. the cost of countermeasure outweighs the value of the asset and potential loss.
    D. The likelihood of the risk occurring is unknown.

95. Which of the following is the **MOST** important management signoff for migrating an order processing system from a test environment to a production environment?

    A. User
    B. Security
    C. Operations
    D. Database

96. The **PRIMARY** objective of a risk management program is to:

    A. minimize inherent risk.
    B. eliminate business risk.
    C. implement effective controls.
    D. reduce residual risk to acceptable levels.

97. The decision on whether new risks should fall under periodic or event-driven reporting should be based on which of the following?

    A. Mitigating controls
    B. Visibility of impact
    C. Likelihood of occurrence
    D. Incident frequency

98. The goals of information security risk management inside an enterprise are **BEST** achieved if these risk management activities are:

    A. treated as a distinct process.
    B. conducted by the IT department.
    C. integrated within business processes.
    D. communicated to all employees.

99. When designing the technical solution for a disaster recovery site, the **PRIMARY** factor that should be taken into consideration is the:

    A. services delivery objective.
    B. recovery time objective (RTO).
    C. recovery window.
    D. maximum tolerable outage (MTO).

100.   It is important to develop an information security baseline because it helps to define:

    A.   critical information resources needing protection.
    B.   a security policy for the entire organization.
    C.   the minimum acceptable security to be implemented.
    D.   required physical and logical access controls.

101.   What is the **PRIMARY** role of the information security manager in the process of information classification within an organization?

    A.   Defining and ratifying the classification structure of information assets
    B.   Deciding the classification levels applied to the organization's information assets
    C.   Securing information assets in accordance with their classification
    D.   Checking if information assets have been classified properly

102.   Which of the following is the **MOST** effective way to treat a risk such as a natural disaster that has a low probability and a high impact level?

    A.   Implement countermeasures.
    B.   Eliminate the risk.
    C.   Transfer the risk.
    D.   Accept the risk.

103.   What is the **MOST** important item to be included in an information security policy?

    A.   The definition of roles and responsibilities
    B.   The scope of the security program
    C.   The key objectives of the security program
    D.   Reference to procedures and standards of the security program

104.   When creating a forensic image of a hard drive, which of the following should be the **FIRST** step?

    A.   Identify a recognized forensics software tool to create the image.
    B.   Establish a chain of custody log.
    C.   Connect the hard drive to a write blocker.
    D.   Generate a cryptographic hash of the hard drive contents.

105.   The **MOST** effective approach to address issues that arise between IT management, business units and security management when implementing a new security strategy is for the information security manager to:

    A.   escalate issues to an external third party for resolution.
    B.   ensure that senior management provide authority for security to address the issues.
    C.   insist that managers or units not in agreement with the security solution accept the risk.
    D.   refer the issues to senior management along with any security recommendations.

106.   Nonrepudiation can **BEST** be ensured by using:

    A.   strong passwords.
    B.   a digital hash.
    C.   symmetric encryption.
    D.   digital signatures.

107. The "separation of duties" principle is violated if which of the following individuals has update rights to the database access control list (ACL)?

    A. Data owner
    B. Data custodian
    C. Systems programmer
    D. Security administrator

108. After a risk assessment, it is determined that the cost to mitigate the risk is much greater than the benefit to be derived. The information security manager should recommend to business management that the risk be:

    A. transferred.
    B. treated.
    C. accepted.
    D. terminated.

109. The **PRIMARY** objective of security awareness is to:

    A. ensure that security policies are understood.
    B. influence employee behavior.
    C. ensure legal and regulatory compliance.
    D. notify of actions for noncompliance.

110. When calculating an annual loss expectancy (ALE), which variable **MOST** requires the information systems (IS) manager to form an opinion based on the uncertainty of the future?

    A. Exposure factor
    B. Asset value
    C. Annual rate of occurrence
    D. Recovery time objective (RTO)

111. The **MOST** appropriate owner of customer data stored in a central database, used only by an organization's sales department, would be the:

    A. sales department.
    B. database administrator.
    C. chief information officer (CIO).
    D. head of the sales department.

112. A risk assessment and business impact analysis (BIA) have been completed for a major proposed purchase and new process for an organization. There is disagreement between the information security manager and the business department manager who will own the process regarding the results and the assigned risk. Which of the following would be the **BEST** approach of the information security manager?

    A. Acceptance of the business manager's decision on the risk to the corporation
    B. Acceptance of the information security manager's decision on the risk to the corporation
    C. Review of the assessment with executive management for final input
    D. A new risk assessment and BIA are needed to resolve the disagreement

113. To determine how a security breach occurred on the corporate network, a security manager looks at the logs of various devices. Which of the following **BEST** facilitates the correlation and review of these logs?

    A. Database server
    B. Domain name server (DNS)
    C. Time server
    D. Proxy server

114.   Which of the following is the **MOST** appropriate method for deploying operating system (OS) patches to production application servers?

     A.   Batch patches into frequent server updates
     B.   Initially load the patches on a test machine
     C.   Set up servers to automatically download patches
     D.   Automatically push all patches to the servers

115.   Which of the following is the **BEST** mechanism to determine the effectiveness of the incident response process?

     A.   Incident response metrics
     B.   Periodic auditing of the incident response process
     C.   Action recording and review
     D.   Postincident review

116.   Which of the following would be the **BEST** way to improve employee attitude toward and commitment to information security?

     A.   Implement restrictive controls.
     B.   Customize methods training to the audience.
     C.   Apply administrative penalties.
     D.   Initiate stronger supervision.

117.   What is the **BEST** way to ensure data protection upon termination of employment?

     A.   Retrieve identification badge and card keys
     B.   Retrieve all personal computer equipment
     C.   Erase all of the employee's folders
     D.   Ensure all logical access is removed

118.   Detailed business continuity plans should be based **PRIMARILY** on:

     A.   consideration of different alternatives.
     B.   the solution that is least expensive.
     C.   strategies that cover all applications.
     D.   strategies validated by senior management.

119.   When performing an information risk analysis, an information security manager should **FIRST**:

     A.   establish the ownership of assets.
     B.   evaluate the risks to the assets.
     C.   take an asset inventory.
     D.   categorize the assets.

120.   An information security manager believes that a network file server was compromised by a hacker. Which of the following should be the **FIRST** action taken?

     A.   Ensure that critical data on the server are backed up.
     B.   Shut down the compromised server.
     C.   Initiate the incident response process.
     D.   Shut down the network.

121.   An information security manager at a global organization has to ensure that the local information security program will initially ensure compliance with the:

    A.   corporate data privacy policy.
    B.   data privacy policy where data are collected.
    C.   data privacy policy of the headquarters' country.
    D.   data privacy directive applicable globally.

122.   Data owners are **PRIMARILY** responsible for:

    A.   providing access to systems.
    B.   approving access to systems.
    C.   establishing authorization and authentication.
    D.   handling identity management.

123.   The use of insurance is an example of which of the following?

    A.   Risk mitigation
    B.   Risk acceptance
    C.   Risk elimination
    D.   Risk transfer

124.   Which of the following is the **BEST** approach to obtain senior management commitment to the information security program?

    A.   Describe the reduction of risk.
    B.   Present the emerging threat environment.
    C.   Benchmark against other enterprises.
    D.   Demonstrate the alignment of the program to business objectives.

125.   Of the following, which is the **MOST** effective way to measure strategic alignment of an information security program?

    A.   Track audits over time.
    B.   Evaluate incident losses.
    C.   Analyze business cases.
    D.   Interview business owners.

126.   An enterprise is transferring its IT operations to an offshore location. An information security manager should be **PRIMARILY** concerned about:

    A.   reviewing new laws and regulations.
    B.   updating operational procedures.
    C.   validating staff qualifications.
    D.   conducting a risk assessment.

127.   Which of the following types of information would the information security manager expect to have the **LOWEST** level of security protection in a publicly traded, multinational enterprise?

    A.   Strategic business plan
    B.   Upcoming financial results
    C.   Customer personal information
    D.   Previous financial results

128. A project manager is developing a developer portal and requests that the security manager assign a public IP address so that it can be accessed by in-house staff and by external consultants outside the organization's local are network (LAN). What should the security manager do **FIRST**?

    A. Understand the business requirements of the developer portal
    B. Perform a vulnerability assessment of the developer portal
    C. Install an intrusion detection system (IDS)
    D. Obtain a signed nondisclosure agreement (NDA) from the external consultants before allowing external access to the server

129. The director of auditing has recommended a specific information security monitoring solution to the information security manager. What should the information security manager do **FIRST**?

    A. Obtain comparative pricing bids and complete the transaction with the vendor offering the best deal.
    B. Add the purchase to the budget during the next budget preparation cycle to account for costs.
    C. Perform an assessment to determine correlation with business goals and objectives.
    D. Form a project team to plan the implementation.

130. Documented standards/procedures for the use of cryptography across the enterprise should **PRIMARILY**:

    A. define the circumstances where cryptography should be used.
    B. define cryptographic algorithms and key lengths.
    C. describe handling procedures of cryptographic keys.
    D. establish the use of cryptographic solutions.

131. A serious vulnerability is reported in the firewall software used by an organization. Which of the following should be the immediate action of the information security manager?

    A. Ensure that all OS patches are up-to-date
    B. Block inbound traffic until a suitable solution is found
    C Obtain guidance from the firewall manufacturer
    D. Commission a penetration test

132. Which of the following is the **MOST** appropriate frequency for updating antivirus signature files for antivirus software on production servers?

    A. Daily
    B. Weekly
    C. Concurrently with O/S patch updates
    D. During scheduled change control updates

133. Risk assessments should be repeated at regular intervals because:

    A. business threats are constantly changing.
    B. omissions in earlier assessments can be addressed.
    C. repetitive assessments allow various methodologies.
    D. they help raise awareness on security in the business.

134. Which of the following should be determined **FIRST** when establishing a business continuity program?

    A. Cost to rebuild information processing facilities
    B. Incremental daily cost of the unavailability of systems
    C. Location and cost of offsite recovery facilities
    D. Composition and mission of individual recovery teams

135.   Previously accepted risk should be:

    A. reassessed periodically since the risk can be escalated to an unacceptable level due to changing conditions.

    B. accepted permanently since management has already spent resources (time and labor) to conclude that the risk level is acceptable.

    C. avoided next time since risk avoidance provides the best protection to the company.

    D. removed from the risk log once it is accepted.

136.   What would a security manager **PRIMARILY** utilize when proposing the implementation of a security solution?

    A. Risk assessment report

    B. Technical evaluation report

    C. Business case

    D. Budgetary requirements

137.   An information security manager performing a security review determines that compliance with access control policies to the data center is inconsistent across employees. The **FIRST** step to address this issue should be to:

    A. assess the risk of noncompliance.

    B. initiate security awareness training.

    C. prepare a status report for management.

    D. increase compliance enforcement.

138.   When implementing security controls, an information security manager must **PRIMARILY** focus on:

    A. minimizing operational impacts.

    B. eliminating all vulnerabilities.

    C. usage by similar organizations.

    D. certification from a third party.

139.   Which of the following technologies is utilized to ensure that an individual connecting to a corporate internal network over the Internet is not an intruder masquerading as an authorized user?

    A. Intrusion detection system (IDS)

    B. IP address packet filtering

    C. Two-factor authentication

    D. Embedded digital signature

140.   Who should determine the appropriate classification of accounting ledger data located on a database server and maintained by a database administrator in the IT department?

    A. Database administrator (DBA)

    B. Finance department management

    C. Information security manager

    D. IT department management

141.   Which of the following would **BEST** prepare an information security manager for regulatory reviews?

    A. Assign an information security administrator as regulatory liaison

    B. Perform self-assessments using regulatory guidelines and reports

    C. Assess previous regulatory reports with process owners input

    D. Ensure all regulatory inquiries are sanctioned by the legal department

142. Who is responsible for ensuring that information is classified?

    A. Senior management
    B. Security manager
    C. Data owner
    D. Custodian

143. When performing a risk assessment, the **MOST** important consideration is that:

    A. management supports risk mitigation efforts.
    B. annual loss expectations (ALEs) have been calculated for critical assets.
    C. assets have been identified and appropriately valued.
    D. attack motives, means and opportunities be understood.

144. Which of the following would be the **FIRST** step when developing a business case for an information security investment?

    A. Defining the objectives
    B. Calculating the cost
    C. Defining the need
    D. Analyzing the cost-effectiveness

145. Which of the following is the **MOST** important item to consider when evaluating products to monitor security across the enterprise?

    A. Ease of installation
    B. Product documentation
    C. Available support
    D. System overhead

146. A database was compromised by guessing the password for a shared administrative account and confidential customer information was stolen. The information security manager was able to detect this breach by analyzing which of the following?

    A. Invalid logon attempts
    B. Write access violations
    C. Concurrent logons
    D. Firewall logs

147. In which phase of the development process should risk assessment be **FIRST** introduced?

    A. Programming
    B. Specification
    C. User testing
    D. Feasibility

148. A contract has just been signed with a new vendor to manage IT support services. Which of the following tasks should the information security manager ensure is performed **NEXT**?

    A. Establish vendor monitoring.
    B. Define reporting relationships.
    C. Create a service level agreement (SLA).
    D. Have the vendor sign a nondisclosure agreement (NDA).

149.   The cost of implementing a security control should not exceed the:

        A.   annualized loss expectancy.
        B.   cost of an incident.
        C.   asset value.
        D.   implementation opportunity costs.

150.   The **PRIMARY** goal of a corporate risk management program is to ensure that an organization's:

        A.   IT assets in key business functions are protected.
        B.   business risks are addressed by preventive controls.
        C.   stated objectives are achievable.
        D.   IT facilities and systems are always available.

151.   Which of the following would **BEST** protect an organization's confidential data stored on a laptop computer from unauthorized access?

        A.   Strong authentication by password
        B.   Encrypted hard drives
        C.   Multifactor authentication procedures
        D.   Network-based data backup

152.   There is reason to believe that a recently modified web application has allowed unauthorized access. Which is the **BEST** way to identify an application backdoor?

        A.   Black box pen test
        B.   Security audit
        C.   Source code review
        D.   Vulnerability scan

153.   It is important to classify and determine relative sensitivity of assets to ensure that:

        A.   cost of protection is in proportion to sensitivity.
        B.   highly sensitive assets are protected.
        C.   cost of controls is minimized.
        D.   countermeasures are proportional to risk.

154.   Temporarily deactivating some monitoring processes, even if supported by an acceptance of operational risk, may not be acceptable to the information security manager if:

        A.   it implies compliance risks.
        B.   short-term impact cannot be determined.
        C.   it violates industry security practices.
        D.   changes in the roles matrix cannot be detected.

155.   Quantitative risk analysis is **MOST** appropriate when assessment data:

        A.   include customer perceptions.
        B.   contain percentage estimates.
        C.   do not contain specific details.
        D.   contain subjective information.

156. An organization's operations staff places payment files in a shared network folder and then the disbursement staff picks up the files for payment processing. This manual intervention will be automated some months later, thus cost-efficient controls are sought to protect against file alterations. Which of the following would be the **BEST** solution?

    A. Design a training program for the staff involved to heighten information security awareness
    B. Set role-based access permissions on the shared folder
    C. The end user develops a PC macro program to compare sender and recipient file contents
    D. Shared folder operators sign an agreement to pledge not to commit fraudulent activities

157. Information security managers should use risk assessment techniques to:

    A. justify selection of risk mitigation strategies.
    B. maximize the return on investment (ROI).
    C. provide documentation for auditors and regulators.
    D. quantify risks that would otherwise be subjective.

158. Which of the following are seldom changed in response to technological changes?

    A. Standards
    B. Procedures
    C. Policies
    D. Guidelines

159. A risk assessment should **TYPICALLY** be conducted:

    A. once a year for each business process and subprocess.
    B. every three to six months for critical business processes.
    C. by external parties to maintain objectivity.
    D. annually or whenever there is a significant change.

160. A third party was engaged to develop a business application. Which of the following is the **BEST** test for the existence of back doors?

    A. System monitoring for traffic on network ports
    B. Security code reviews for the entire application
    C. Reverse engineering the application binaries
    D. Running the application from a high-privileged account on a test system

161. Segregation of duties assists with:

    A. employee monitoring.
    B. reduced supervisory requirements.
    C. fraud prevention.
    D. enhanced compliance.

162. For risk management purposes, the value of a physical asset should be based on:

    A. original cost.
    B. net cash flow.
    C. net present value.
    D. replacement cost.

163.   A multinational organization operating in fifteen countries is considering implementing an information security program. Which factor will **MOST** influence the design of the Information security program?

    A. Representation by regional business leaders
    B. Composition of the board
    C. Cultures of the different countries
    D. IT security skills

164.   Which of the following would be **MOST** appropriate for collecting and preserving evidence?

    A. Encrypted hard drives
    B. Generic audit software
    C. Proven forensic processes
    D. Log correlation software

165.   Which of the following has the highest priority when defining an emergency response plan?

    A. Critical data
    B. Critical infrastructure
    C. Safety of personnel
    D. Vital records

166.   The **BEST** way to standardize security configurations in similar devices is through the use of:

    A. policies.
    B. procedures.
    C. technical guides.
    D. baselines.

167.   Risk assessment is **MOST** effective when performed:

    A. at the beginning of security program development.
    B. on a continuous basis.
    C. while developing the business case for the security program.
    D. during the business change process.

168.   Which of the following is the **MOST** important aspect of forensic investigations that will potentially involve legal action?

    A. The independence of the investigator
    B. Timely intervention
    C. Identifying the perpetrator
    D. Chain of custody

169.   Which of the following **BEST** describes the scope of risk analysis?

    A   Key financial systems
    B. Organizational activities
    C. Key systems and infrastructure
    D. Systems subject to regulatory compliance

170.   Successful implementation of information security governance will **FIRST** require:

    A. security awareness training.
    B. updated security policies.
    C. a computer incident management team.
    D. a security architecture.

171.   An organization's information security strategy should be based on:

    A. managing risk relative to business objectives.
    B. managing risk to a zero level and minimizing insurance premiums.
    C. avoiding occurrence of risks so that insurance is not required.
    D. transferring most risks to insurers and saving on control costs.

172.   Which of the following would be **MOST** helpful to achieve alignment between information security and organization objectives?

    A. Key control monitoring
    B. A robust security awareness program
    C. A security program that enables business activities
    D. An effective security architecture

173.   Which of the following should be determined while defining risk management strategies?

    A. Risk assessment criteria
    B. Organizational objectives and risk appetite
    C. IT architecture complexity
    D. Enterprise disaster recovery plans

174.   Which of the following is the **MOST** important step before implementing a security policy?

    A. Communicating to employees
    B. Training IT staff
    C. Identifying relevant technologies for automation
    D. Obtaining sign-off from stakeholders

175.   From an information security manager perspective, what is the immediate benefit of clearly-defined roles and responsibilities?

    A. Enhanced policy compliance
    B. Improved procedure flows
    C. Segregation of duties
    D. Better accountability

176.   The **PRIMARY** objective for information security program development should be:

    A. establishing strategic alignment with the business.
    B. establishing incident response programs.
    C. identifying and implementing the best security solutions.
    D. reducing the impact of the risk in the business.

177.   The **MOST** important reason for conducting periodic risk assessments is because:

    A.   risk assessments are not always precise.
    B.   security risks are subject to frequent change.
    C.   reviewers can optimize and reduce the cost of controls.
    D.   it demonstrates to senior management that the security function can add value.

178.   What is the **BEST** method for mitigating against network denial of service (DoS) attacks?

    A.   Ensure all servers are up-to-date on OS patches
    B.   Employ packet filtering to drop suspect packets
    C.   Implement network address translation to make internal addresses nonroutable
    D.   Implement load balancing for Internet facing devices

179.   The **MOST** appropriate individual to determine the level of information security needed for a specific business application is the:

    A.   system developer.
    B.   information security manager.
    C.   steering committee.
    D.   system data owner.

180.   The implementation of continuous monitoring controls is the **BEST** option where:

    A.   incidents may have a high impact and frequency
    B.   legislation requires strong information security controls
    C.   incidents may have a high impact but low frequency
    D.   electronic commerce is a primary business driver

181.   Which of the following is **MOST** effective in protecting against the attack technique known as phishing?

    A.   Firewall blocking rules
    B.   Up-to-date signature files
    C.   Security awareness training
    D.   Intrusion detection monitoring

182.   Which of the following is an example of a corrective control?

    A.   Diverting incoming traffic as a response to a denial of service (DoS) attack
    B.   Filtering network traffic
    C.   Examining inbound network traffic for viruses
    D.   Logging inbound network traffic

183.   The **MOST** effective way to incorporate risk management practices into existing production systems is through:

    A.   policy development.
    B.   change management.
    C.   awareness training.
    D.   regular monitoring.

184.   Which of the following is an indicator of effective governance?

    A.   A defined information security architecture
    B.   Compliance with international security standards
    C.   Periodic external audits
    D.   An established risk management program

185.   Which of the following is the **BEST** way to mitigate the risk of the database administrator reading sensitive data from the database?

    A.   Log all access to sensitive data.
    B.   Employ application-level encryption.
    C.   Install a database monitoring solution.
    D.   Develop a data security policy.

186.   When securing wireless access points, which of the following controls would **BEST** assure confidentiality?

    A.   Implementing wireless intrusion prevention systems
    B.   Not broadcasting the service set IDentifier (SSID)
    C.   Implementing wired equivalent privacy (WEP) authentication
    D.   Enforcing a virtual private network (VPN) over wireless

187.   An information security manager has implemented procedures for monitoring specific activities on the network. The system administrator has been trained to analyze the network events, take appropriate action and provide reports to the information security manager. What additional monitoring should be implemented to give a more accurate, risk-based view of network activity?

    A.   The system administrator should be monitored by a separate reviewer.
    B.   All activity on the network should be monitored.
    C.   No additional monitoring is needed in this situation.
    D.   Monitoring should be done only by the information security manager.

188.   To **BEST** improve the alignment of the information security objectives in an organization, the chief information security officer (CISO) should:

    A.   revise the information security program.
    B.   evaluate a balanced business scorecard.
    C.   conduct regular user awareness sessions.
    D.   perform penetration tests.

189.   Because of its importance to the business, an organization wants to quickly implement a technical solution which deviates from the company's policies. An information security manager should:

    A.   conduct a risk assessment and allow or disallow based on the outcome.
    B.   recommend a risk assessment and implementation only if the residual risks are accepted.
    C.   recommend against implementation because it violates the company's policies.
    D.   recommend revision of current policy.

190.   Which of the following is the **MOST** serious exposure of automatically updating virus signature files on every desktop each Friday at 11:00 p.m. (23.00 hrs.)?

    A.   Most new viruses' signatures are identified over weekends
    B.   Technical personnel are not available to support the operation
    C.   Systems are vulnerable to new viruses during the intervening week
    D.   The update's success or failure is not known until Monday

191. Which of the following devices should be placed within a DMZ?

    A. Router
    B. Firewall
    C. Mail relay
    D. Authentication server

192. Information security policy enforcement is the responsibility of the:

    A. security steering committee.
    B. chief information officer (CIO).
    C. chief information security officer (CISO).
    D. chief compliance officer (CCO).

193. What is the **PRIMARY** objective of a post-event review in incident response?

    A. Adjust budget provisioning
    B. Preserve forensic data
    C. Improve the response process
    D. Ensure the incident is fully documented

194. Effective IT governance is **BEST** ensured by:

    A. utilizing a bottom-up approach.
    B. management by the IT department.
    C. referring the matter to the organization's legal department.
    D. utilizing a top-down approach.

195. Investments in information security technologies should be based on:

    A. vulnerability assessments.
    B. value analysis.
    C. business climate.
    D. audit recommendations.

196. Which of the following is the **MOST** important objective of an information security strategy review?

    A. Ensuring that risks are identified and mitigated
    B. Ensuring that information security strategy is aligned with organizational goals
    C. Maximizing the return of information security investments
    D. Ensuring the efficient utilization of information security resources

197. Which of the following would be the **MOST** important factor to be considered in the loss of mobile equipment with unencrypted data?

    A. Disclosure of personal information
    B. Sufficient coverage of the insurance policy for accidental losses
    C. Potential impact of the data loss
    D. Replacement cost of the equipment

198.   An organization's board of directors has learned of recent legislation requiring organizations within the industry to enact specific safeguards to protect confidential customer information. What actions should the board take next?

     A.   Direct information security on what they need to do
     B.   Research solutions to determine the proper solutions
     C.   Require management to report on compliance
     D.   Nothing; information security does not report to the board

199.   Risk acceptance is a component of which of the following?

     A.   Assessment
     B.   Treatment
     C.   Evaluation
     D.   Monitoring

200.   Who is ultimately responsible for ensuring that information is categorized and that protective measures are taken?

     A.   Information security officer
     B.   Security steering committee
     C.   Data owner
     D.   Data custodian

Page intentionally left blank

CISM® Review Questions, Answers & Explanations Manual 2012

SAMPLE EXAM ANSWER AND REFERENCE KEY

Exam Question Number	ANSWER	REF.	Exam Question Number	ANSWER	REF.	Exam Question Number	ANSWER	REF.	Exam Question Number	ANSWER	REF.	Exam Question Number	ANSWER	REF.
1	C	S3-254	51	B	S1-11	101	A	S1-52	151	B	S3-73
2	D	S2-85	52	D	S2-28	102	C	S2-102	152	C	S3-232
3	C	S2-38	53	B	S1-46	103	C	S1-134	153	D	S2-36
4	B	S1-25	54	C	S4-13	104	B	S4-88	154	A	S2-134
5	A	S4-46	55	B	S2-17	105	D	S1-74	155	B	S2-18
6	A	S1-15	56	C	S4-3	106	D	S3-90	156	B	S3-221
7	D	S2-37	57	A	S4-99	107	C	S3-192	157	A	S2-32
8	C	S4-27	58	A	S2-9	108	C	S2-83	158	C	S1-22
9	C	S2-115	59	C	S2-61	109	B	S3-131	159	D	S2-15
10	D	S1-51	60	B	S3-187	110	C	S2-112	160	B	S3-200
11	B	S4-102	61	B	S1-50	111	D	S2-86	161	C	S2-116
12	B	S2-47	62	A	S4-10	112	C	S1-100	162	D	S2-6
13	A	S3-47	63	D	S2-19	113	C	S4-56	163	C	S1-78
14	D	S2-129	64	D	S3-154	114	B	S3-147	164	C	S4-81
15	A	S1-93	65	A	S3-259	115	D	S4-72	165	C	S4-65
16	B	S1-66	66	C	S3-55	116	B	S3-244	166	D	S2-118
17	C	S2-46	67	C	S1-40	117	D	S3-189	167	B	S2-88
18	C	S4-91	68	D	S3-116	118	D	S4-36	168	D	S4-82
19	A	S4-85	69	B	S3-22	119	C	S2-44	169	B	S2-22
20	A	S2-100	70	C	S1-81	120	C	S4-78	170	B	S1-10
21	C	S1-39	71	A	S1-91	121	B	S1-44	171	A	S1-90
22	A	S2-23	72	A	S2-50	122	B	S1-122	172	C	S1-82
23	D	S4-93	73	C	S3-246	123	D	S2-124	173	B	S1-107
24	D	S2-41	74	D	S4-75	124	D	S1-126	174	D	S3-91
25	B	S2-1	75	C	S2-16	125	D	S1-112	175	D	S1-38
26	A	S4-34	76	B	S4-100	126	D	S2-121	176	D	S1-123
27	B	S3-139	77	D	S3-115	127	D	S2-55	177	B	S2-2
28	D	S3-151	78	B	S2-78	128	A	S2-62	178	B	S4-52
29	D	S2-60	79	D	S2-56	129	C	S1-111	179	D	S3-101
30	A	S4-70	80	C	S2-14	130	A	S3-157	180	A	S3-199
31	D	S1-94	81	B	S3-135	131	C	S4-58	181	C	S3-44
32	D	S3-265	82	D	S1-71	132	A	S3-17	182	A	S4-55
33	C	S2-25	83	A	S4-92	133	A	S2-49	183	B	S2-10
34	B	S4-9	84	D	S3-203	134	B	S4-1	184	D	S1-119
35	D	S1-87	85	C	S1-24	135	A	S2-106	185	B	S3-253
36	A	S4-45	86	B	S2-63	136	C	S1-63	186	D	S3-261
37	C	S3-242	87	B	S3-84	137	A	S2-126	187	A	S3-278
38	B	S4-44	88	C	S2-111	138	A	S2-103	188	B	S3-61
39	A	S3-119	89	D	S1-5	139	C	S3-29	189	B	S2-70
40	D	S4-94	90	C	S2-73	140	B	S3-177	190	C	S4-26
41	B	S1-55	91	C	S3-233	141	B	S1-37	191	C	S3-5
42	C	S1-20	92	B	S2-128	142	C	S2-82	192	C	S1-42
43	A	S1-76	93	D	S3-144	143	C	S2-74	193	C	S4-35
44	C	S2-57	94	C	S2-54	144	C	S1-117	194	D	S1-96
45	B	S2-53	95	A	S3-110	145	D	S3-39	195	B	S1-7
46	C	S3-41	96	D	S2-42	146	A	S4-54	196	B	S1-125
47	D	S2-105	97	B	S2-12	147	D	S2-52	197	C	S2-77
48	B	S3-188	98	C	S2-114	148	A	S3-267	198	C	S1-102
49	A	S4-22	99	C	S4-43	149	C	S1-13	199	B	S2-13
50	B	S1-53	100	C	S3-58	150	C	S2-35	200	B	S3-160

Reference example: S1-132 = See domain 1, question 132 for explanation of answer.

Page intentionally left blank

CISM® Review Questions, Answers & Explanations Manual 2012

SAMPLE EXAM ANSWER SHEET (PRETEST)

(side 1)

Please use this answer sheet to take the sample exam as a pretest to determine strengths and weaknesses. The answer key/reference grid is on page 225.

(side 2)

Please use this answer sheet to take the sample exam as a pretest to determine strengths and weaknesses. The answer key/reference grid is on page 225.

YOUR SIGNATURE/SEAL REQUIRED HERE: _____

Mark Reflex® by NCS EM-238649-1:654321   HR04   Printed in U.S.A.   © Copyright 2001 by National Computer Systems, Inc. All rights reserved.

SAMPLE

Chicago is:
1. a country
2. a mountain
3. an island
4. a city

WRONG ⊙
WRONG ⊘
WRONG ⊗
RIGHT ●

CISM® Review Questions, Answers & Explanations Manual 2012
SAMPLE EXAM ANSWER SHEET (POSTTEST)

(side 1)

Please use this answer sheet to take the sample exam as a posttest to determine strengths and weaknesses. The answer key/reference grid is on page 225.

(side 2)

Please use this answer sheet to take the sample exam as a posttest to determine strengths and weaknesses. The answer key/reference grid is on page 225.

YOUR SIGNATURE/SEAL REQUIRED HERE:

	A	B	C	D		A	B	C	D		A	B	C	D		A	B	C	D		A	B	C	D
81	○	○	○	○	101	○	○	○	○	121	○	○	○	○	141	○	○	○	○	161	○	○	○	○	181	○	○	○	○
82	○	○	○	○	102	○	○	○	○	122	○	○	○	○	142	○	○	○	○	162	○	○	○	○	182	○	○	○	○
83	○	○	○	○	103	○	○	○	○	123	○	○	○	○	143	○	○	○	○	163	○	○	○	○	183	○	○	○	○
84	○	○	○	○	104	○	○	○	○	124	○	○	○	○	144	○	○	○	○	164	○	○	○	○	184	○	○	○	○
85	○	○	○	○	105	○	○	○	○	125	○	○	○	○	145	○	○	○	○	165	○	○	○	○	185	○	○	○	○
86	○	○	○	○	106	○	○	○	○	126	○	○	○	○	146	○	○	○	○	166	○	○	○	○	186	○	○	○	○
87	○	○	○	○	107	○	○	○	○	127	○	○	○	○	147	○	○	○	○	167	○	○	○	○	187	○	○	○	○
88	○	○	○	○	108	○	○	○	○	128	○	○	○	○	148	○	○	○	○	168	○	○	○	○	188	○	○	○	○
89	○	○	○	○	109	○	○	○	○	129	○	○	○	○	149	○	○	○	○	169	○	○	○	○	189	○	○	○	○
90	○	○	○	○	110	○	○	○	○	130	○	○	○	○	150	○	○	○	○	170	○	○	○	○	190	○	○	○	○
91	○	○	○	○	111	○	○	○	○	131	○	○	○	○	151	○	○	○	○	171	○	○	○	○	191	○	○	○	○
92	○	○	○	○	112	○	○	○	○	132	○	○	○	○	152	○	○	○	○	172	○	○	○	○	192	○	○	○	○
93	○	○	○	○	113	○	○	○	○	133	○	○	○	○	153	○	○	○	○	173	○	○	○	○	193	○	○	○	○
94	○	○	○	○	114	○	○	○	○	134	○	○	○	○	154	○	○	○	○	174	○	○	○	○	194	○	○	○	○
95	○	○	○	○	115	○	○	○	○	135	○	○	○	○	155	○	○	○	○	175	○	○	○	○	195	○	○	○	○
96	○	○	○	○	116	○	○	○	○	136	○	○	○	○	156	○	○	○	○	176	○	○	○	○	196	○	○	○	○
97	○	○	○	○	117	○	○	○	○	137	○	○	○	○	157	○	○	○	○	177	○	○	○	○	197	○	○	○	○
98	○	○	○	○	118	○	○	○	○	138	○	○	○	○	158	○	○	○	○	178	○	○	○	○	198	○	○	○	○
99	○	○	○	○	119	○	○	○	○	139	○	○	○	○	159	○	○	○	○	179	○	○	○	○	199	○	○	○	○
100	○	○	○	○	120	○	○	○	○	140	○	○	○	○	160	○	○	○	○	180	○	○	○	○	200	○	○	○	○

Mark Reflex® by NCS EM-238649-1:654321   HR04   Printed in U.S.A.   © Copyright 2001 by National Computer Systems, Inc. All rights reserved.

SAMPLE

Chicago is:
1. a country
2. a mountain
3. an island
4. a city

WRONG ○○○ WRONG ○○○
WRONG ○○○ RIGHT ○○●
WRONG ○⊘○

EVALUATION

ISACA continuously monitors the swift and profound professional, technological and environmental advances affecting IS audit and control professionals. Recognizing these rapid advances, CISM review manuals are updated annually.

To assist ISACA in keeping abreast of these advances, please take a moment to evaluate the *CISM® Review Questions, Answers & Explanations Manual 2012*. Such feedback is valuable to fully serve the profession and future CISM exam registrants.

To complete the evaluation on the web site, please go to *www.isaca.org/studyaidsevaluation*.

Thank you for your support and assistance.

Prepare for the 2012 CISM Exams

2012 CISM Review Resources for Exam Preparation and Professional Development

Successful Certified Information Security Manager® (CISM®) exam candidates have an organized plan of study. To assist individuals with the development of a successful study plan, ISACA® offers several study aids and review courses to exam candidates. These include:

Study Aids

- *CISM® Review Manual 2012*

- *CISM® Review Questions, Answers & Explanations Manual 2012*

- *CISM® Review Questions, Answers & Explanations Manual 2012 Supplement*

- CISM® Practice Question Database v12

To order, visit www.isaca.org/cismbooks.

Review Courses

- Chapter-sponsored review courses

To find or register for a course in your region, visit *www.isaca.org/cismreview*.

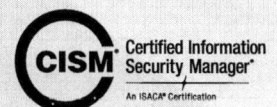

HISACA®
Trust in, and value from, information systems

CISM Certified Information Security Manager®
An ISACA® Certification

CISM Review Questions, Answers & Explanations Manual 2012
ISACA. All Rights Reserved.